Michele Aresta, Angela Dibenedetto, Franck Dumeignil (Eds.)
Cyclic Economy

Carbon-Cyclic Economy

Edited by
Michele Aresta, Angela Dibenedetto, Franck Dumeignil

Cyclic Economy

Policies in Major Countries and the EU

Edited by
Michele Aresta, Angela Dibenedetto, Franck Dumeignil

DE GRUYTER

Editors
Prof. Michele Aresta
Innovative Catalysis for Carbon Recycling
IC^{2}R Ltd
Via Casamassima km 3
70010 Valenzano (BA)
Italy
michele.aresta@ic2r.com

Prof. Angela Dibenedetto
Department of Chemistry
University of Bari and CIRCC
via Orabona 4
70126 Bari
Italy
angela.dibenedetto@uniba.it

Prof. Franck Dumeignil
Unité de Catalyse et Chimie du Solide
University Lille
C3 Batiment
59655 Villeneuve d'Ascq Cedex
France
franck.dumeignil@univ-lille.fr

ISBN 978-3-11-076707-0
e-ISBN (PDF) 978-3-11-076717-9
e-ISBN (EPUB) 978-3-11-076719-3

Library of Congress Control Number: 2024937889

Bibliographic information published by the Deutsche Nationalbibliothek
The Deutsche Nationalbibliothek lists this publication in the Deutsche Nationalbibliografie;
detailed bibliographic data are available on the internet at http://dnb.dnb.de.

© 2024 Walter de Gruyter GmbH, Berlin/Boston
Cover image: enot-poloskun/iStock/Getty Images Plus
Typesetting: Integra Software Services Pvt. Ltd.

www.degruyter.com

Preface

This book is the first of a series dedicated to "circular economy," an economic model that will characterize our society in the coming years. With the advent of the "industrialization," humans slowly moved away from agriculture as the main economic activity towards industrial production. They also left behind the "nature-inspired" (and, thus, circular) production and use of goods and moved towards the "use once and throw away" behavior typical of the "linear economy" model based on the false assumptions that natural resources are illimited, waste can be carelessly disposed, and our planet can buffer any impact. Two hundred years of linear economy have evidenced that natural resources are finite and the planet has limited capacity of recovery from anthropic burdens.

Our society needs to re-introduce nature-inspired behaviors for avoiding serious impacts on the delicate equilibria which maintain life on our planet. It is believed that the continued increase of anthropogenic CO_2 emissions (37 Gt/y these days), and its accumulation in the atmosphere is causing "climate change." As a matter of fact, the natural C-cycle, which turns over 750 Gt/y of CO_2, is not able to buffer the non-natural emissions (amounting to less than 5% of the natural cycle). This is only one example.

The impact of human activities on the natural compartments (air, soil, water) is growing with the growth of the population and their demand of conforts. Therefore, humans must adapt their lifestyles to ensure earth remains a viable habitat, while also ensuring that future generations have access to sufficient resources. Our use of goods today is reckless: using products only once, we shorten their life and reduce their value. Recycling, which is typical of nature, must enter into our attitude at all levels, so to extend the life and increase the value of goods (Figure 1).

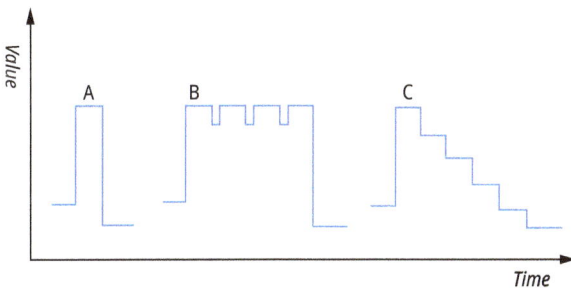

Figure 1: Re-use (B, C) extends the life and increases the value of goods with respect to single use (A).

Recycling of goods gives them several lives, either at the same value (Figure 1, B) or at lower values (Figure 1, C). Case B is typical of recycling metals, while case C is frequent with plastics and paper. Thermoplastics used for food applications will most likely be recycled as plastics for lower-value applications (packaging, wrapping, tubes, and similar) due to the impossibility of recovering the original purity. Similarly, waste white

https://doi.org/10.1515/9783110767179-202

paper used for documents will not recover its original white color and strength because of the difficulty to eliminate colors and inks without affecting the physical properties of fibers. Noteworthy, plastics could in principle be brought to their original quality status by depolymerizing them and repolymerizing monomers. For paper, the full requalification is prevented today by the fact that chemical and physical treatments reduce the fibers resistance. Recycled paper may recover the desired color, but the quality of the paper will be affected so that anyway the second-generation paper will have a lower value.

Recycling is not a new attitude. Man is recycling metals and water since long time. Table 1 gives an idea of where we are today with recycling goods.

Table 1: Recycling of goods today.

Good	Worldwide used volume/ mass	% Recycling average	Notes
Freshwater (global: agricultural, industrial and drinking)	3,996,757,700,000 cubic meters/y Per capita: largest user Turkmenistan 16,281 L/d; lowest: DR Congo 34 L/d. Average use: 70% agriculture, 20% industry, 10% drinking. In industrialized countries, the industrial use of water can reach 80% (Belgium)	<1% A target of 30% worldwide average recycling is set for 2050.	Recycling best cases: Israel 85%; Kuwait 35%; Singapore 30%; Queensland (AUS): >35%; California (USA): 50% in Orange county.
Paper	423 Mt/y (2017)	*Ca.* 50% Recycling paper saves 65% of the energy needed to make new paper and reduces water pollution by 35% and air pollution by 74%.	Recycling one ton of paper saves up to 31 trees, 4,000 kWh of energy, 1.7 barrels (270 liters) of oil, 10.2 million Btu's of energy, 26,000 liters of water and 3.5 cubic meters of landfill space.
Iron	1,000 Mt/y	40%	Iron is the good recycled since longer time.
Aluminum	*ca.* 94 Mt/y (production 2017) *ca.* 81 Mt/y (used 2017)	31 In special cases, Al is made of 75% recycled material.	Aluminum is one of the most commonly recycled goods.
Copper	19.1 Mt/y (2017)	>45%	Copper can be recycled without loss of properties.
Plastics	360 Mt/y	9%	Recycling depends on the composition and structure of the polymeric material.

Recycling carbon for implementing a *carbon-cyclic economy* is a great challenge for science and technology. Our Society has to shift from fossil carbon (see the COP28 agreement, unanimously signed by 197 countries and the EU) and, thus, needs alternative primary carbon for applications (chemical industry, avio-fuels) that still need it. Sources of alternative-C have been identified into: recycled plastics, waste biomass and CO_2, each with its own limits. The chemical industry emits some 930 Mt/y of CO_2 and uses some 230 Mt/y of CO_2 to produce chemicals. Using CO_2 as a source of carbon for producing non-fossil fuels (e-fuels, solar fuels) would enlarge the amount of CO_2 used by at least one order of magnitude, cutting equivalent amounts of fossil carbon. Process innovation based on recycling is already under progress and early changes can be noted in some manufacturing processes: Industry is getting cleaner. Results are under our eyes for the textile, construction, chemical, steel, and IT sectors. The reduction of emissions (gases, liquids, solids) can be performed via a process-system organization and a procurement strategy that may render waste of a process a useful raw material for another.

The EU Green Deal and the coming Directives on Industrial Carbon Management (ICM) will put a strong and defined frame to the innovation in industrial production within the EU. Member States have to adopt wise regulations and legislations to implement the changes.

This book presents the strategies some of the world's major economies intend to apply in the field of the circular economy. The series will be implemented with similar publications covering other major world economies and selected industrial sectors.

Michele Aresta   Angela Dibenedetto   Franck Dumeignil
IC^2R Ltd   University of Bari and CIRCC   University of Lille

Contents

List of contributors

Lazarević Jelisaveta
Metropolitan University
FEFA Faculty

Pitić Goran
Metropolitan University
FEFA Faculty

Vlačić Ernest
Metropolitan University
FEFA Faculty

Michele Galatola
Senior Policy Officer
European Commission
Directorate-General for Internal Market,
Industry, Entrepreneurship and SMEs
(DG GROW)

Pengxian Ye
Department of Chemistry
Faculty of Science
The Chinese University of Hong Kong
Shatin
Hong Kong SAR
China

Debjyoti Ray
Department of Chemistry
Faculty of Science
The Chinese University of Hong Kong
Shatin
Hong Kong SAR
China

Chunshan Song
Department of Chemistry
Faculty of Science
The Chinese University of Hong Kong
Shatin
Hong Kong SAR
China

Kandasamy Palanivelu
Centre for Environmental Studies
And
Centre for Climate Change and Disaster
Management
Department of Civil Engineering
College of Engineering Guindy Campus
Anna University
Chennai – 600025
Tamil Nadu
India

Sathyanarayanan Sri Shalini
Centre for Climate Change and Disaster
Management
Department of Civil Engineering
College of Engineering Guindy Campus
Anna University
Chennai – 600025
Tamil Nadu
India

Aline S. Tavares
Universidade Federal do Rio de Janeiro
Escola de Química
Av. Athos da Silveira Ramos 149
CT bl E
Cidade Universitária
21941-909 Rio de Janeiro
Brazil

Suzana Borschiver
Universidade Federal do Rio de Janeiro
Escola de Química
Av. Athos da Silveira Ramos 149
CT bl E
Cidade Universitária
21941-909 Rio de Janeiro
Brazil

https://doi.org/10.1515/9783110767179-204

Claudio J. A. Mota
Universidade Federal do Rio de Janeiro
Escola de Química
Av. Athos da Silveira Ramos 149
CT bl E
Cidade Universitária
21941-909 Rio de Janeiro
Brazil
And
INCT Energia & Ambiente
UFRJ
21941-909 Rio de Janeiro
Brazil

Kibeak Lee
Carbon Neutral Strategy Center
Korea Research Institute of Chemical Technology
(KRICT)
141 Gajeong-ro
Yuseong-gu
Daejeon
Republic of South Korea

Lazarević Jelisaveta, Pitić Goran, and Vlačić Ernest

1 From linear to circular: creating and delivering sustainable value

Abstract: The green transition fosters the emergence of circular economy firms committed to sustainable practices and resource optimization. In what ways can the business formulate organizational operations, business models, products and materials that not only eradicate waste and pollution but also advance sustainability? This book chapter endeavors to elucidate the evolution of an economy and business models that concurrently benefit both enterprises and the natural environment. By comprehensively exploring the transition from linear to circular paradigms, it intricately examines the process of envisioning a sustainable global economy. Moreover, the chapter underscores the pivotal role of emerging technologies in catalyzing and shaping a circular future. Given the substantial influence of businesses in shaping a circular economy, this section accentuates innovative approaches to value creation, delivery, and capture that are in harmony with sustainability principles.

1.1 Circular economy as an arising megatrend

We are currently experiencing times characterized by heightened levels of volatility, uncertainty, complexity, and ambiguity (VUCA) unlike any other period in history [1]. These profound shifts in our civilization, known as "civilizational trends," are often referred to as "global megatrends" (Figure 1.1) due to their widespread impact on a global scale. Consulting firm Roland Berger [2] forecasts similar megatrends continuing up to 2050, emphasizing the centrality of technology, environmental concerns, and humanity in addressing contemporary civilizational challenges.

A recent report [Circularity gap report, 3] reveals that circularity has attained megatrend status, evidenced by a nearly threefold increase in discussions, reports, and articles on the concept since 2018. Concurrently, from 2018 onward, the Human Development Index has ascended across high-, middle-, and low-income nations, correlating with increased material consumption and material footprint per capita. The report highlights that humans have consumed more than 500 gigatones of material, representing a significant 28% portion of total material consumption since 1900.

Transitioning to the twenty-first century, an examination of the Global Price Commodity Index reveals a remarkable trend: over a mere two decades, commodity prices soared by 153% in 2023 compared to 2003 (Figure 1.2). This rapid escalation under-

Lazarević Jelisaveta, Pitić Goran, and Vlačić Ernest, Metropolitan University, FEFA Faculty

https://doi.org/10.1515/9783110767179-001

The Roland Berger Trend Compendium 2050 covers 6 Megatrends
That shapes the future development of our wold until 2050

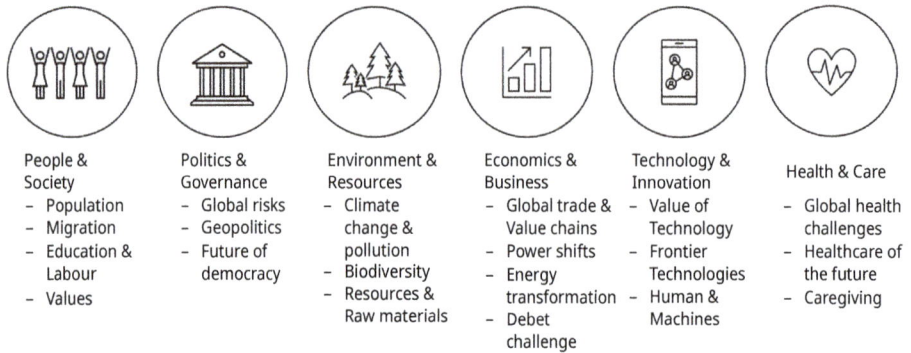

People & Society	Politics & Governance	Environment & Resources	Economics & Business	Technology & Innovation	Health & Care
– Population – Migration – Education & Labour – Values	– Global risks – Geopolitics – Future of democracy	– Climate change & pollution – Biodiversity – Resources & Raw materials	– Global trade & Value chains – Power shifts – Energy transformation – Debt challenge	– Value of Technology – Frontier Technologies – Human & Machines	– Global health challenges – Healthcare of the future – Caregiving

Figure 1.1: Global megatrends.
Source: Roland Berger

scores the evolving dynamics within global markets and the need for proactive strategies to navigate the complexities of a rapidly changing economic landscape.

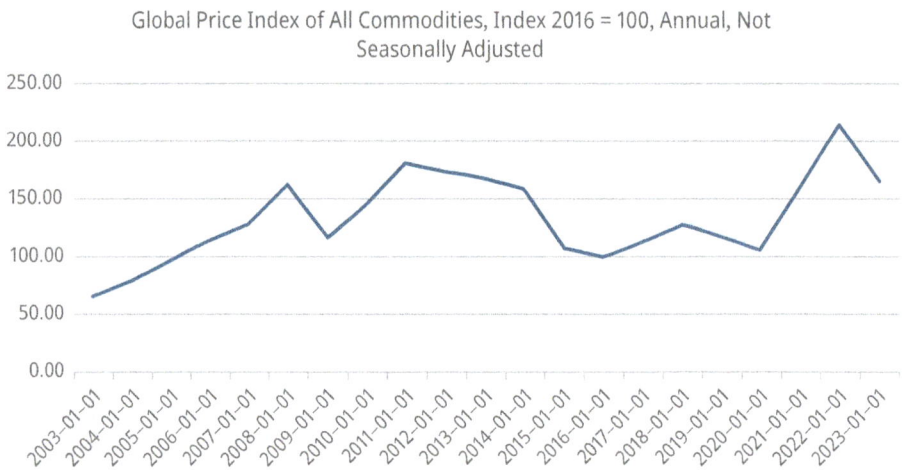

Figure 1.2: Global price index of all commodities.
Source: International Monetary Fund, Global Price Index of All Commodities [PALLFNFINDEXQ], retrieved from FRED, Federal Reserve Bank of St. Louis; https://fred.stlouisfed.org/series/PALLFNFINDEXQ, March 2024.

In one of its reports, the Elen Macarthur Foundation explains that as the global middle class is poised to more than double its current size, there will be a significant increase in consumption and demand for materials. This surge is expected to drive up hardly predictable input costs and lead to price fluctuations, posing particular challenges as accessing new resources becomes progressively more difficult and expensive [4]. Given these circumstances, both businesses and governments require a new industrial model – one that is less reliant on primary energy and material inputs [5]. What's needed is a business model that fosters resilience and hedging, enabling the generation of revenue that is less dependent on the volatility of input costs.

Under mentioned circumstances, this chapter elucidates the evolution of a circular economy (CE), an economy, and business models that concurrently benefit both enterprises and the natural environment. By comprehensively exploring the transition from linear to circular paradigms, it intricately examines the process of envisioning a sustainable global economy. Moreover, the chapter underscores the pivotal role of emerging technologies in catalyzing and shaping a circular future. Given the substantial influence of businesses in shaping a CE, this section accentuates innovative approaches to value creation, delivery, and capture that are in harmony with sustainability principles.

1.2 Circular economy phenomena

Since the Industrial Revolution, our economies have been grounded in the fundamental characteristics of a linear economy, where available natural resources are used and processed by labor and energy to produce products, thrown away after use, as waste. However, the mentioned challenges and trends in the global economy are putting pressure to change the current take-make-waste system.

Various research studies indicate that the current model of economic growth contributes to environmental degradation. Given the reality that there is "no planet B," the CE approach has garnered increasing attention as it aims to reshape traditional patterns of consumption and production.

The notion of a CE and its integration into industrial processes trace back to the late 1970s [6]. The industrial implementation of the CE was pioneered by Stahel and Ready in 1976 [7]. Before this period, in the 1960s, Professor Kenneth E. Boulding first developed the idea of the CE, conceptualizing the Earth as a closed system and advocating for the coexistence of the economy and the environment in balance [7, 8]. Aligned, the concept of CE is primarily grounded in ecological and environmental economics and industrial ecology [9]. In the past decade, this concept has emerged as a compelling alternative to the prevailing economic model of "take, make, and dispose" [10].

One of the most renewed definitions of CE is provided by the Elen Macarthur Foundation, which describes it as "a system where materials never become waste and nature is regenerated. In a CE, products and materials are kept in circulation through processes like maintenance, reuse, refurbishment, remanufacture, recycling, and composting." [35] However, it should be noted that there is no static definition of CE [11], and several schools of thought shape it.

One aspect is associated with regenerative design [12], which revolves around the concept that all processes and systems can renew and regenerate the resources they utilize and consume [6]. The other school of thought is related to the performance economy, introduced by Stahel and Ready who outlined industrial strategies aimed at waste prevention and the promotion of regional job creation [7]. Stahel also stated that product-life extension reduces the usage of natural resources and increases wealth, while presenting the way toward transition to a more sustainable society [13]. Moreover, the CE approach is altering the paradigm of product usage, thereby influencing existing business models for delivering value to the market. It advocates for a concept wherein the manufacturer or seller retains ownership of the product, offering a service for its use rather than selling the product itself [6]. Stahel was among the pioneers who emphasized this concept as one of the most significant business models in the loop/CE [13].

If we apply the cyclical processes found in nature to the industrial ecosystem, we are coming to the next school of thought that refers to "cradle-to-cradle" [14]. It adopts the productive processes of natural or biological metabolism as a model for developing technical metabolism. This approach enables the design of product components for continuous recovery [6].

From its inception, the concept of the CE has been closely linked with research in industrial ecology [15, 11], which seeks to connect various actors within an industrial ecosystem to create circular processes where waste serves as an input. This brings us to the concept of the blue economy [16], wherein waste transforms into a resource for generating new revenue, with particular emphasis placed on integrating local communities into the entire process.

To simplify the concept of a CE, we can draw parallels with basic knowledge taught in primary school. In nature, waste is nearly nonexistent as materials flow seamlessly between organisms, with one species' waste becoming another's sustenance, all powered by the sun. Conversely, in industrial settings, we follow a linear model of "make, take, and dispose," leading to waste generation, especially of nonbiodegradable materials. The question arises: Can we emulate the efficiency of natural systems in industry and extract value from waste?

Dealing with nonbiodegradable materials involves creating products from cycled materials – materials that can be reused to produce renewed products that are resold. This approach is aligned with the product-life extension process beyond its expiration date. A step closer to a CE is achieved by separating product ingredients – biological components safely returned to nature and technical components designed for reuse. Furthermore, if this entire process operates on renewable energy and companies exert positive pressure on suppliers to join this circular approach, we have the economy functioning on circular principles. According to Elon McArthur Foundation [6], those principles refer to:

- When both technical and biomaterials are created to cycle, waste is designed out. Biological ones can be composted, and the technical ones are created to be reused.
- Natural systems are adaptive, whereas our linear systems, which are designed in complete opposition to how nature functions, are less resilient.
  Just as in financial markets, where investing solely in individual stocks exposes us to the risk of one company, we can mitigate risk by diversifying our portfolio and investing in multiple stocks. Similarly, diversifying the business value creation itself can enhance our resilience.
- A glimpse into natural systems shows that the sun serves as the primary source of energy. To establish a CE, we should power it with renewable energy. By revisiting the initial graph and contemplating methods to hedge against volatile prices and economic shocks, powering the system with renewable sources can render it less vulnerable and more resilient.
- No business, nor the governments are isolated from the broader systems. And every action in those systems has its reaction and spillover effects on other aspects of the ecosystem. The same principle applies in the CE, to make the system circular leaders of certain fields should take into consideration all actors and their influence.
- At the end, we circle back to the concept of waste and the principle of transitioning from technical to biological nutrients, with a cascade approach at the core of the CE (Figure 1.3).

As illustrated in the preceding graph, in the CE, products crafted from technical materials undergo a transition from being consumed to being utilized by the user. This transition creates a wide array of opportunities for the development of new consumption and sales models. Furthermore, these new business models are bolstered by the development of emerging technologies. The emergence of new technologies is anticipated to bring about more significant changes than those witnessed in the past 100 years. Digital technologies are already integrated into products and services, generating data that is then utilized to enhance business performance and provide value to the market. The CE system is expected to further enhance information feedback loops and profoundly alter the approach to value creation within the circular model [5]. Furthermore, the CE is deeply intertwined with significant technological trends, including climate technologies, renewable energy, and the transition to a net-zero economy. Additionally, it serves as a prominent application domain for emerging technologies such as applied artificial intelligence (AI), the Internet of Things (IoT), and industrial machine learning. In the context of the Fifth Industrial Revolution, which prioritizes human–machine interactions and the well-being of societies and customers, the concept of the CE is gaining momentum as an integral component of broader environmental, social, and governance (ESG) initiatives.

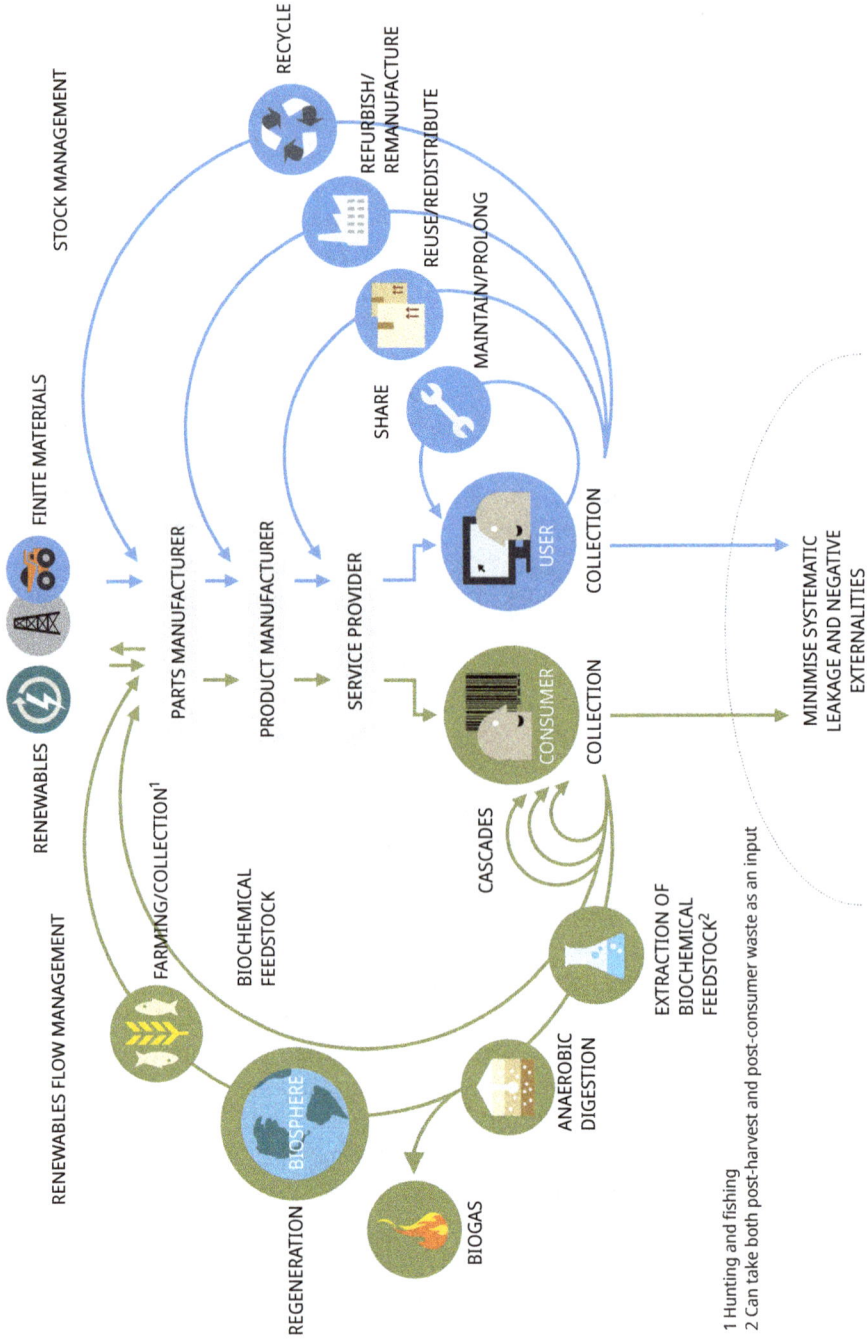

Figure 1.3: The brief diagram of the circular economy.
Source: Ellen MacArthur Foundation [5]

However, global circularity has declined from 9.1% in 2018 to 7.2% in 2023 (measured as a share of secondary materials, The Circularity Gap Report, 2024). With this 21% decrease over five years, consumption continues to accelerate. This underscores the reality that while the proportion of secondary materials is decreasing, the overall quantity of consumed materials is increasing. In the past six years alone, the global economy has consumed more than half a trillion units of materials, approaching the total amount used throughout the entire twentieth century (The Circularity Gap Report, 2024).

These statistics highlight a sobering reality: despite the megatrend status of the CE, the global economy still lacks models that effectively translate the concept into tangible realization and practical implementation. The CE faces ongoing challenges in breaking through barriers to seize new circular opportunities. Addressing this trend requires the development of new technologies and business models that untangle supply chain complexities, foster collaboration, and deliver sustainable value.

1.3 Role of state-of-the-art technologies in advancing the circular economy

We are currently living in the era of the Fourth Industrial Revolution and rapidly approaching to the era of the fifth one. In such technological and civilization transitions as global challenges related to resource depletion and environmental degradation intensify, the CE emerges as a powerful solution. CE aims to decouple economic growth from resource consumption by emphasizing reuse, recycling, and responsible production [17]. To achieve this, we must harness the potential of advanced technologies.

1.3.1 Digital technologies driven transition

The transition toward a CE necessitates innovative approaches that blend cutting-edge technologies with sustainable practices, leading us to state-of-the-art (SOA) digital technologies that contribute to the proliferation of circularity, thus becoming greentech or cleantech. Since their impact on environmental and economic performance, these technologies play a crucial role in reshaping supply chains and promoting sustainable resource use [18].

The fusion of SOA technologies with the principles of the CE represents a pioneering approach to environmental sustainability and economic resilience. This synergy is not merely about integrating technology into current systems but about reimagining and redesigning our industrial and economic models from the ground up. By leveraging cutting-edge innovations [19, 20], we can transition toward a more sustainable, efficient, and inclusive CE.

Technologies such as AI, the IoT, blockchain, big data analytics, 5G, high-performance computing (HPC), and augmented and virtual realities (AR/VR) play important roles in this transformation. These technologies and innovations based on them hold immense promise for advancing circularity [17].

Here are some examples:

- AI and machine learning can predict the lifespan of products, optimize resource use, and significantly enhance recycling processes. Thus, AI can sort recyclable materials more accurately and at a faster rate than human workers, increasing the efficiency of recycling centers [21]. AI-driven platforms can facilitate the matching of waste producers with recycling operators, optimizing the recycling value chain.
- IoT devices enable real-time tracking and monitoring of resources, providing valuable data for optimizing product use and material flows. IoT sensors can monitor the condition of products and components, identifying when they can be repaired or need to be recycled, thus extending their lifecycle [22].
- Blockchain technology offers a secure and transparent way to track the lifecycle of products and materials, ensuring traceability and accountability in the supply chain. It can support the certification of recycled materials and the verification of sustainable practices, building trust among consumers and stakeholders [23].

Currently, the dominating industrial technology platform, Industry 4.0, is about making manufacturing smarter through real-time data, machine learning, and interconnectedness, which can lead to more efficient, flexible, and customizable production processes. The incoming one, Industry 5.0 adds two crucial components to the I4.0, which enable the CE to elevate to the next level of human centricity and sustainability.

Aiming to envision what might be the appropriate fit in their interaction, a conceptual matrix to align Industry 5.0 technologies with the challenges encountered in various segments of the CE is presented in Table 1.1.

This matrix demonstrates the potential of the Industry 5.0 technology platform aiming to revolutionize the CE by addressing its core challenges. It's important to recognize that the synergy between these technologies and circular principles can lead to innovative solutions that not only reduce environmental impact but also offer new business opportunities and models for sustainable growth, described in the chapter that follows.

Digital platforms for CE facilitate the exchange, sharing, and repair of goods, promoting the utilization of idle assets and extending product lifecycles [7]. Platforms such as online marketplaces for second-hand goods or services for sharing resources among businesses can significantly reduce waste and promote the efficient use of resources [24].

Table 1.1: Industry 5.0 technologies answering challenges in circular economy.

CE segment	Challenge	Industry 5.0 technology	How it addresses the challenge
Resource recovery	Efficient recycling and waste management	Robotics and automation	Advanced sorting technologies and robotics can increase the efficiency and precision of recycling processes.
		AI and machine earning	AI can optimize waste processing and identify materials that can be recovered more effectively.
Product design	Design for longevity, repairability, and reuse	3D printing/ additive manufacturing	Enables the production of complex parts that are easier to repair or recycle, and use of recycled materials in new products. SOA manufacturing technologies enable efficient resource utilization. Through precision engineering, additive manufacturing, and smart production systems, we can maximize the lifespan of products and minimize waste. For instance, 3D printing allows customized designs, reducing material excess and energy consumption.
		IoT	Enables products to communicate their status, facilitating maintenance and efficient end-of-life management.
Supply chain and reverse logistics	Transparency and traceability	Blockchain	Provides a secure and transparent record of product origins, material use, and recycling paths. Digital tools enhance supply chain visibility and traceability. Blockchain ensures transparent transactions, while real-time data analytics optimize material flows. By minimizing inefficiencies and reducing transportation distances, these technologies contribute to circular supply chains.
		IoT	Real-time tracking of materials and products to ensure optimal logistics and reduce waste.
		smart manufacturing	Digital solutions facilitate reverse logistics – the process of reclaiming, refurbishing, and recycling products. Smart sensors track product lifecycles, enabling efficient returns and remanufacturing. This closed-loop approach reduces the need for virgin resources.

Table 1.1 (continued)

CE segment	Challenge	Industry 5.0 technology	How it addresses the challenge
Business models	Promoting product-as-a-service (PaaS)	AI and big data	Enables predictive maintenance and usage-based services, enhancing the PaaS model.
		Cloud computing	Facilitates the sharing of resources and services, supporting collaborative consumption models.
Consumer engagement	Encouraging sustainable consumption	AR/VR	Provides immersive experiences to educate consumers about sustainability and product lifecycle.
		Mobile and web apps	Engages consumers in circular practices like sharing, leasing, and recycling through easy-to-use platforms.

Source: Authors' presentation.

1.3.2 Digital is not enough

The proliferation of the CE requires a comprehensive technological approach that extends beyond digital innovations. While digital technologies are pivotal in this transformation, the role of SOA technologies extends beyond the digital realm, encompassing biotechnologies, advanced materials, and novel manufacturing processes that collectively drive the proliferation of the CE. Their synergy with digital technologies amplifies their potential in the CE.

1.3.2.1 Advanced materials: the building blocks of CE

Advanced materials, characterized by their superior properties such as strength, durability, and lightness, play a crucial role in extending the lifespan of products and facilitating recycling and reuse. Biomaterials, for example, offer sustainable alternatives to fossil-based materials, reducing the environmental footprint of products from production to disposal [25]. Innovative materials like self-healing concrete and biodegradable plastics not only enhance product longevity but also ensure that end-of-life disposal aligns with ecological principles, thus supporting the CE model.

1.3.2.2 Biotechnologies: closing the loop

Biotechnologies, including enzyme-based recycling and bio-based production processes, offer transformative approaches to waste management and material recovery in the CE. Enzymatic recycling of plastics, as explored by Tournier et al. [26], demonstrates a promising method for breaking down PET plastics into their monomers, which can then be reused to produce new plastics without loss of quality. This biological cycle mimics natural processes, ensuring that products can be fully reintegrated into the biosphere or technosphere, thereby closing the loop.

1.3.2.3 Novel manufacturing processes: rethinking production

Novel manufacturing processes such as 3D printing and modular construction significantly contribute to resource efficiency and waste reduction. 3D printing minimizes material waste through additive manufacturing, allowing for the precise use of materials and the customization of products to exact specifications [27]. Moreover, modular construction promotes the reuse and refurbishment of components in the building sector, extending the lifecycle of materials and reducing construction waste.

This multifaceted approach not only addresses the environmental challenges of our time but also offers new opportunities for innovation, competitiveness, and growth. The future of the CE lies in the integration of these SOA technologies, marking a new era of environmental stewardship and economic prosperity.

1.3.3 Controversial use of technology in boosting CE

Using waste as a source of energy, often referred to as "waste-to-energy" (WtE), is a controversial and complex issue that sits at the intersection of waste management and renewable energy production. While WtE presents a solution to mitigate landfill usage and generate energy, it is not without its drawbacks and contradictions. Although WtE is often touted as a renewable energy source, it is not entirely carbon-neutral. The combustion of waste materials, especially those containing fossil-based plastics, releases carbon dioxide, a greenhouse gas. Consequently, reliance on WtE can contribute to carbon emissions, conflicting with global efforts to reduce greenhouse gas levels and combat climate change [28].

The economic viability of WtE technologies is another area of debate. The initial capital costs for establishing WtE plants are high, and their economic performance is sensitive to changes in energy prices and the availability of waste feedstock. These economic uncertainties can make WtE projects less attractive to investors and policymakers compared to other renewable energy sources or waste management strategies [29]. While WtE can play a role in managing waste and generating energy, it is essential to carefully consider its environmental, economic, and societal impacts. A holistic approach to waste management, prioritizing waste reduction, reuse, and recycling before energy recovery, is crucial for sustainable development.

1.3.4 Startup scene a catalyzer of innovation

Besides large corporations and academia, small and medium-sized enterprises (SMEs), and startups are at the forefront of innovating within the CE, demonstrating agility and creativity on par with larger corporations. These smaller entities often exhibit greater flexibility in adopting new business models and technological innovations that promote resource efficiency, waste reduction, and sustainable product design [30]. This dynamism enables SMEs and startups to quickly respond to emerging trends and challenges in the CE, contributing significantly to its proliferation and success strongly advancing GREENTECH and CLEANTECH. By their nature, startups are agile and innovative, often capable of rapidly developing and deploying new technologies that can disrupt traditional industries. In the realm of greentech and cleantech, startups have the potential to introduce groundbreaking solutions that reduce waste, improve resource efficiency, and facilitate the transition from a linear to a circular economic model. For example, the development of biodegradable materials, energy-efficient manufacturing processes, and platforms for the sharing economy are areas where startups have already begun to make significant impacts [7]. These innovations not only contribute to environmental sustainability but also offer new business models that can be economically beneficial. By fostering a startup ecosystem focused on sustainability, we can accelerate the proliferation of the CE, making it a more viable and attractive option for businesses and consumers alike.

1.3.5 Boosting circular economy with SOA technology use cases

State-of-the-art (SOA) technologies have increasingly become catalysts for disruptive innovation in the CE, offering groundbreaking solutions that redefine waste management, resource efficiency, and sustainable production. Here are several compelling use cases where these technologies have made significant impacts:

– **Enzymatic plastic recycling:** Companies like Carbios in France have developed an enzymatic process to depolymerize PET plastic waste into its original monomers, which can then be used to produce new PET plastics of virgin quality. This innovation disrupts traditional mechanical recycling methods by enabling infinite recycling of plastics without quality degradation, significantly contributing to the CE.
– **Modular building techniques:** KieranTimberlake, an architectural firm, has leveraged modular construction techniques to create buildings that can be easily disassembled and reused. Their approach allows for components and materials to

be efficiently recycled and repurposed, reducing waste and promoting sustainability in the construction industry.

– **Blockchain for supply chain transparency:** The provenance platform utilizes blockchain technology to enhance transparency across global supply chains, enabling businesses and consumers to trace the origins and lifecycle of products. This visibility fosters responsible sourcing, waste reduction, and adherence to CE principles by highlighting sustainable practices and encouraging their adoption.

– **AI-driven material discovery for CE:** Companies like Material Connexion use AI to accelerate the discovery and development of sustainable and recyclable materials. By harnessing vast databases and machine learning algorithms, they can identify materials that meet specific criteria for sustainability and circularity, speeding up innovation and application in various industries.

– **3D printing with recycled materials:** Startups like Reflow have created a sustainable business model around using recycled plastics to produce 3D printing filament. This approach not only provides a high-quality, sustainable input for 3D printing enthusiasts and professionals but also demonstrates a lucrative method of upcycling plastic waste.

– **Circular fashion platforms:** The RealReal and Vestiaire Collective, luxury consignment online retailers, have revolutionized the fashion industry by creating platforms for buying and selling preowned luxury goods. Their business models encourage the reuse and prolonged lifecycle of high-quality garments, reducing waste and promoting sustainability in the fashion sector.

– **Smart asset management for electronics:** Dutch startup, Closing the Loop, offers a "waste-compensation" service for electronics, where they recycle an equivalent amount of electronic waste for every device that a company purchases. This innovative approach turns the linear consumption of electronics into a circular model, ensuring that the environmental impact of new devices is offset by responsibly recycling an equal amount of e-waste.

Presented use cases demonstrate the potential of SOA technologies to drive significant advancements in the CE, showcasing how innovative solutions can lead to more sustainable and efficient practices across various sectors.

1.4 Circular business models

The concept of circular business models (CBMs) has been developed on the notions of CE and business model innovations. Schwager and Moser [31] have introduced the idea of developing new business models to encourage a shift from selling products to supplying services with positive outcomes for the economy and the environment. Since then, the literature has grown exponentially on this topic by raising awareness of the relevance of this theme for the future of both business and mankind among major stakeholders. Mentink [32] explains CBM by using a standard construe of a business model that creates, delivers, and captures value, but with and within closed material loops. Geissdoerfer et al. [33] describe CBM as "business models that are cycling, extending, intensifying, and/or dematerializing material and energy loops to reduce the resource inputs into and the waste and emission leakage out of an organizational system," which leads us to the four strategies for CBM often found in the literature (Figure 1.4).

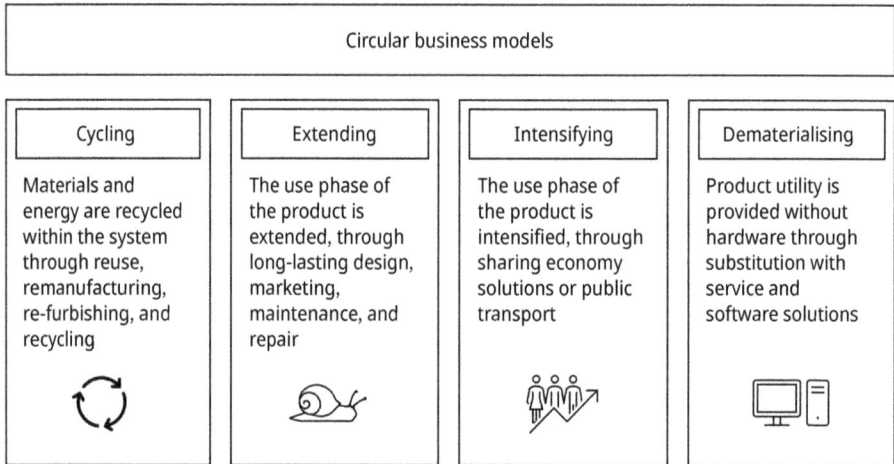

Circular business models			
Cycling	**Extending**	**Intensifying**	**Dematerialising**
Materials and energy are recycled within the system through reuse, remanufacturing, re-furbishing, and recycling	The use phase of the product is extended, through long-lasting design, marketing, maintenance, and repair	The use phase of the product is intensified, through sharing economy solutions or public transport	Product utility is provided without hardware through substitution with service and software solutions

Figure 1.4: Circular business models.
Source: Geissdoerfer et al. [33]

Key components and characteristics of CBMs encompass various elements that distinguish them from traditional linear models. CBMs prioritize resource optimization by minimizing waste and maximizing value retention throughout the product lifecycle. They aim to create closed-loop systems where materials are continuously circulated within the economy through reuse, recycling, remanufacturing, or regeneration (instead of take–use–make–waste). Products are designed with circularity in mind, emphasizing durability, reparability, modularity, and recyclability. Many circular models shift from selling products to providing services, such as leasing or sharing, which prolongs product lifespan and promotes resource efficiency. Effective management of product returns and end-of-life processes, including collection, refurbishment, recycling, and responsible disposal, is essential for circularity. CBM often involves collaboration and partnerships across the value chain, fostering innovation, resource sharing, and knowledge exchange. Circular models adopt a lifecycle perspective, considering environmental, social, and economic impacts across all stages, from raw material extraction to end-of-life disposal or regeneration. Integration of circular principles across the entire value chain, including suppliers, manufacturers, distributors, retailers, and customers, is crucial for maximizing circularity. Beyond minimizing harm, circular models aim to contribute positively to environmental and social systems, fostering regeneration, biodiversity, and community resilience. Embracing a culture of continuous learning, innovation, and adaptation, CBMs encourage experimentation and evolution toward greater circularity. Companies implementing circular models prioritize transparency and accountability, providing stakeholders with clear information about their environmental performance, progress toward circular goals, and social impacts.

A CBM, unlike a traditional linear value chain, prioritizes circularity as the primary driver of value creation, rather than treating it as an ancillary aspect. To integrate circularity into the heart of value creation, organizations must (re)configure their products and services, their relationships with suppliers and consumers, and their internal operations. Only then can circularity become the cornerstone of the firm's value creation, rather than merely a cost-saving measure or waste management practice.

1.4.1 The circular value creation

Adopting CBMs offers a plethora of benefits that align with economic, environmental, and social priorities. Circularity addresses resource scarcity by minimizing extraction and enhancing efficiency, ultimately aiming to reduce waste and pollution to promote environmental preservation. Moreover, transitioning to circularity yields economic advantages, as it can lead to cost savings through enhanced resource management and decreased waste disposal expenses. Additionally, businesses embracing circularity differentiate themselves in the marketplace, appealing to environmentally conscious consumers and gaining a competitive edge, while circular models aid in regulatory compliance, mitigate risks, and foster innovation. They strengthen customer loyalty and brand reputation while meeting stakeholder expectations for sustainability. Circular practices contribute to long-term sustainability and resilience, supporting global sustainability goals and promoting economic development and job creation. Therefore, embracing circularity is essential for businesses to thrive in a world facing environmental and social challenges.

To create and deliver value through the concept of CE, a whole new approach is needed. These new approaches include concepts like ownership retention by the manufacturer or seller (as proposed by Stahel in 1982) and leverage existing frameworks such as sharing, leasing, and renting for various services. These models have the potential to gain momentum, particularly among newer generations who prioritize access over ownership, as evidenced by trends in the real estate market, shared mobility, and similar solutions.

In a CE, business models provide a broader array of value creation strategies compared to those found in linear value chains. Circular businesses focus extensively on the usage phase of products, generating revenue through service provision rather than product sales. This shift signifies a move away from consumption-centered strategies toward value creation around utilization. Rather than emphasizing a singular use cycle, attention is directed toward resource recovery and the establishment of multiple use cycles to maintain the value of products, materials, and components for as long as feasible. Circular businesses reevaluate traditional producer–consumer relationships, value creation activities, and the configuration of value chains. Environmental and social considerations are integrated into the overarching business culture and philosophy. Ultimately, the goal is to transition away from the linear "take-make-dispose" model toward more regenerative and sustainable ways of doing business.

Value creation in CBMs goes beyond traditional economic measures, encompassing environmental, social, and economic benefits. CBMs optimize resource use, minimize waste generation, and reduce reliance on virgin materials, leading to cost savings in procurement, production, and waste management. By designing products for longevity and reparability, companies decrease lifecycle costs. Diversifying supply chains and embracing circular practices enhance resilience to disruptions and mitigate risks associated with resource scarcity and climate change.

CBMs drive innovation, differentiation, and competitive advantage, attracting customers, investors, and talent. Circular products resonate with environmentally conscious consumers, enhancing brand loyalty and trust. They unlock opportunities for new markets, revenue streams, and market differentiation. Embracing circularity helps companies comply with environmental regulations and access markets with sustainability requirements. Furthermore, CBMs generate benefits for the wider economy while creating employment opportunities, stimulating local economies, and contributing to sustainable development goals. They reduce environmental impacts, preserve natural capital, and enhance ecosystem resilience. CBMs foster collaboration, partnerships, and collective action to address systemic challenges. Adopting circular practices enhances brand reputation, trust, and social license to operate, attracting stakeholders aligned with sustainable business practices. Overall, CBMs align economic prosperity with environmental stewardship and social well-being, driving sustainable and inclusive growth for businesses, communities, and ecosystems.

1.4.2 Variety of circular business models

Circular business models encompass a variety of approaches aimed at achieving sustainability, resource efficiency, and waste reduction. These models, which can be implemented individually or in combination, contribute to advancing the transition toward a more sustainable and regenerative CE.

- Product-as-a-service (PaaS), where companies offer products as services to promote durability and resource optimization.
- Sharing platforms facilitate the sharing or renting of goods to optimize utilization and reduce ownership.

- Remanufacturing and refurbishment involve restoring used products to extend their lifespan.
- Resource recovery and regeneration focus on recovering materials from products and regenerating them into new ones.
- Circular supply chains optimize material flows and minimize waste.
- Product life extension offers services to prolong product life.
- Performance-based contracts tie payment to outcomes or usage rather than ownership.
- Bioeconomy and biomimicry use biological resources and principles.
- Platform cooperatives share ownership and governance among users.
- Integrated ecosystems create closed-loop systems for resource exchange.

Atasu et al. [34] have outlined the circularity matrix (Figure 1.5), where they suggest a combination of three strategies as an optimal way for companies to create CMB for their products (retain product ownership [RPO], product life extension [PLE], and design for recycling [DFR]). The sustainability of CBM depends on the possibility of extracting the economic value from the recovery of the product. They suggest high-value products that are easy to access and process as the strongest candidates for introducing circularity in business models.

Challenges and opportunities are inherent in the adoption and implementation of CBMs, and understanding and addressing these factors are crucial for businesses to effectively transition to circularity. On the challenges side, transitioning from linear to CBMs requires significant organizational change, investment, and coordination across the value chain, which can be particularly challenging for large, established organizations. Additionally, implementing CBMs often requires access to advanced technologies and infrastructure for processes such as recycling and remanufacturing, which may pose barriers to adoption. Moreover, consumer awareness and demand for circular products and services are relatively low, requiring efforts to educate consumers and overcome misconceptions. Regulatory constraints, supply chain complexity, and traceability issues further complicate the adoption of circularity. However, numerous opportunities arise from embracing CBMs. Businesses can differentiate themselves in the marketplace and gain a competitive advantage by appealing to environmentally conscious consumers and investors. Cost savings, efficiency improvements, and innovation opportunities are also prevalent, as circularity fosters the development of new technologies and business models. Furthermore, embracing circularity enhances resilience to supply chain risks, builds stakeholder engagement and reputation, and aligns with global sustainability goals. By addressing challenges and seizing opportunities, businesses can unlock the full potential of CBMs, driving sustainable growth, innovation, and value creation for all stakeholders.

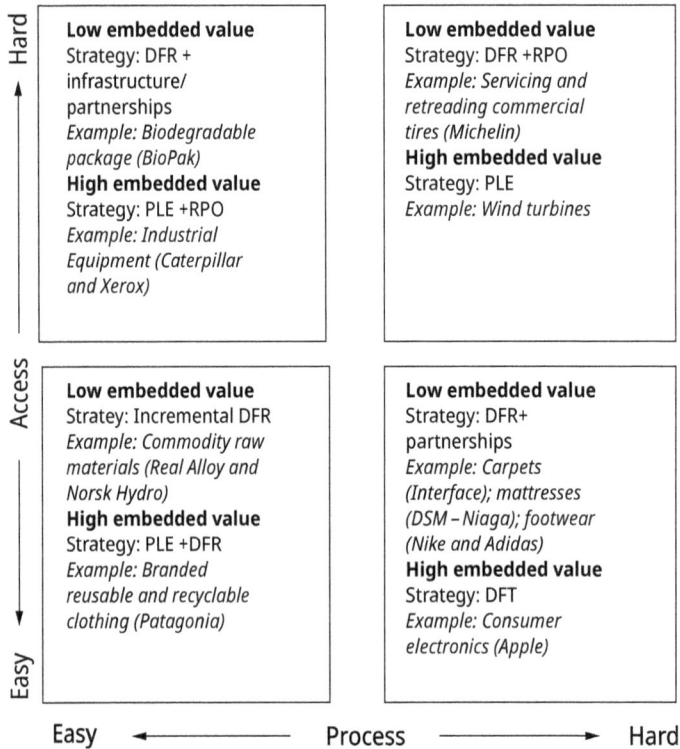

Figure 1.5: The circularity matrix..
Source: Atasu et al. [34]

1.5 Future directions and conclusions

Despite the potential of technologies and innovative business models to accelerate CE, several challenges remain. These include the need for significant investments in technology, the development of standards and regulations to support CE practices, and the transformation of consumer behavior toward more sustainable consumption patterns. Addressing these challenges requires a collaborative effort among governments, businesses, and civil society to create an enabling environment for CE. The faster adoption of SOA technologies and innovative business models holds the key to accelerating the transition to a CE. By leveraging digital technologies such as AI, IoT, and blockchain, and embracing business models that promote sustainability and resource efficiency, society can move toward a more sustainable and resilient future. However, overcoming the existing challenges will require concerted efforts and collaboration across all sectors of the economy and society.

References

[1] Burke, W.W., "Leaders: The strategies for taking charge," by Warren Bennis and Burt Nanus (Book Review). Human Resource Management, 1985, 24(4), 503.

[2] Roland, B., Trend Compendium 2050: Six megatrends will shape the next decades. 2023. Retrieved from: https://www.rolandberger.com/en/Insights/Global-Topics/Trend-Compendium/

[3] Circle Economy Foundation. The circularity gap report. 2024. Retrieved from: https://drive.google.com/file/d/15droT_mBFK6Kkd1aO5kPzYFUqLdul2qM/view

[4] Ellen Macarthur Foundation. Towards the circular economy. Opportunities for the consumer goods sector. Retrieved from: Towards the circular economy Vol. 2: opportunities for the consumer goods sector.pdf (thirdlight.com) 2013b.

[5] Ellen Macarthur Foundation. Towards the circular economy. Accelerating the scale-up across global supply chains. Retrieved from: Towards the circular economy Vol 3: Accelerating the scale-up across global supply chains.pdf (thirdlight.com) 2014.

[6] Ellen, M.F., Towards the circular economy. Economic and business rationale for an accelerated transition. Retrieved from: Towards the circular economy Vol 1: an economic and business rationale for an accelerated transition.pdf (thirdlight.com) 2013a.

[7] Geissdoerfer, M., Savaget, P., Bocken, N.M. and Hultink, E.J., The circular economy–A new sustainability paradigm?. Journal of Cleaner Production, 2017, 143, 757–768.

[8] George, D.A., Lin, B.C.A. and Chen, Y., A circular economy model of economic growth. Environmental Modelling & Software, 2015, 73, 60–63.

[9] Ghisellini, P., Cialani, C. and Ulgiati, S., A review on circular economy: The expected transition to a balanced interplay of environmental and economic systems. Journal of Cleaner Production, 2016, 114, 11–32.

[10] Ness, D., Sustainable urban infrastructure in China: Towards a factor 10 improvement in resource productivity through integrated infrastructure systems. The International Journal of Sustainable Development & World Ecology, 2008, 15(4), 288–301.

[11] Merli, R., Preziosi, M. and Acampora, A., How do scholars approach the circular economy? A systematic literature review. Journal of Cleaner Production, 2018, 178, 703–722.

[12] Lyle, J.T., Regenerative Design for Sustainable Development. John Wiley & Sons, New York; Chichester. 1994.

[13] Stahel, W.R., The product life factor. *An Inquiry into the Nature of Sustainable Societies: The Role of the Private Sector (Series: 1982 Mitchell Prize Papers), NARC,* 1982, 74–96.

[14] Braungart, M., McDonough, W. and Bollinger, A., Cradle-to-cradle design: Creating healthy emissions–a strategy for eco-effective product and system design. Journal of Cleaner Production, 2007, 15(13–14), 1337–1348.

[15] Andersen, M.S., An introductory note on the environmental economics of the circular economy. Sustainability Science, 2007, 2(1), 133–140.

[16] Pauli, G.A., The Blue Economy: 10 Years, 100 Innovations, 100 Million Jobs. Paradigm publications, 2010.

[17] Khan, S.A.R., Piprani, A.Z. and Yu, Z., Digital technology and circular economy practices: Future of supply chains. Operations Management Research, 2022, 15(3), 676–688.

[18] Kolaro, K., Pitić, G., Vlačić, E. and Milosavljević, U., Competitiveness and sustainability in small and open economies in the age of industry 5.0. Ekonomika Preduzeća, 2023, 71(1–2), 113–127.

[19] Dabić, M., Vlačić, E., Ramanathan, U. and Egri, C.P., Evolving absorptive capacity: The mediating role of systematic knowledge management. IEEE Transactions on Engineering Management, 2019, 67(3), 783–793.

[20] Vlačić, E., Dabić, M., Daim, T. and Vlajčić, D., Exploring the impact of the level of absorptive capacity in technology development firms. Technological Forecasting and Social Change, 2019, 138, 166–177.

[21] McKinsey & Company. Artificial intelligence and the circular economy: AI as a tool to accelerate the transition. 2023. Retrieved from https://www.mckinsey.com/capabilities/sustainability/our-insights/artificialintelligence-and-the-circular-economy-ai-as-a-tool-to-accelerate-thetransition.

[22] Gartner,. Top 10 strategic technology trends for 2020. 2019. Retrieved from https://www.gartner.com/smarterwithgartner/gartner-top-10-strategictechnology-trends-for-2020/.

[23] Jabbour, C.J.C., De Sousa Jabbour, A.B.L., Sarkis, J. and Godinho Filho, M., Unlocking the circular economy through new business models based on large-scale data: An integrative framework and research agenda. Technological Forecasting and Social Change, 2019, 144, 546–552.

[24] MDPI. (2023). Chi, Z., Liu, Z., Wang, F. and Osmani, M., Driving circular economy through digital technologies: Current research status and future directions. Sustainability, 2023, 15(24), 16608. https://doi.org/10.3390/su152416608

[25] Kirchherr, J., Reike, D. and Hekkert, M., Conceptualizing the circular economy: An analysis of 114 definitions. Resources, Conservation and Recycling, 2017, 127, 221–232.

[26] Tournier, V., Topham, C.M., Gilles, A., David, B., Folgoas, C., Moya-Leclair, E., Kamionka, E., Desrousseaux, M.L., Texier, H., Gavalda, S. and Cot, M., An engineered PET depolymerase to break down and recycle plastic bottles. Nature, 2020, 580(7802), 216–219.

[27] Ford, S. and Despeisse, M., Additive manufacturing and sustainability: An exploratory study of the advantages and challenges. Journal of Cleaner Production, 2016, 137, 1573–1587.

[28] Arena, U., Process and technological aspects of municipal solid waste gasification. A review. Waste Management, 2012, 32(4), 625–639.

[29] Kumar, A. and Samadder, S.R., A review on technological options of waste to energy for effective management of municipal solid waste. Waste Management, 2017, 69, 407–422.

[30] Rizos, V., Behrens, A., Van der Gaast, W. and Hofman, E., The role of SMEs in green growth and climate change: Lessons from the UNEP/OECD global green growth knowledge platform. Journal of Cleaner Production, 2016, 155, 1–13.

[31] Schwager, P. and Moser, F., The application of chemical leasing business models in Mexico. Environmental Science and Pollution Research, 2006, 13, 131–137. DOI: http://dx.doi.org/10.1065/espr2006.02.294

[32] Mentink, B.A.S., Circular business model innovation: A process framework and a tool for business model innovation in a circular economy. Master thesis. 2014. http://resolver.tudelft.nl/uuid:c2554c91-8aaf-4fdd-91b7-4ca08e8ea621

[33] Geissdoerfer, M., Pieroni, M.P., Pigosso, D.C. and Soufani, K., Circular business models: A review. Journal of Cleaner Production, 2020, 277, 123741. https://doi.org/10.1016/j.jclepro.2020.123741

[34] Atasu, A., et al. The circular business model. Harvard Business Review, 2021. vol July-August

[35] Ellen Macarthur Foundation. What is a circular economy? Retrieved from: https://www.ellenmacarthurfoundation.org/topics/circular-economy-introduction/overview

Michele Galatola

2 Circular economy and product sustainability in Europe

Abstract: This chapter delves into the intricate landscape of the circular economy and product sustainability within Europe, addressing the multifaceted challenges, evolving policies, and the critical roles of various stakeholders in steering towards a more sustainable economic model.

After describing the existing interwoven environmental challenges, such as climate change, biodiversity loss, water scarcity, and land degradation, the author elaborates on the need to rely on innovative approaches in policy-making, with Life Cycle Assessment (LCA) emerging as a fundamental tool.

The author then describes the current State of Product Sustainability, explaining how linear production models are progressively evolving into circular ones. Analyzing the data reveals a mixed picture; while there are some positive signs of decoupling economic growth from resource use, challenges persist, particularly with indirect emissions from imported products. The role of European sustainability agenda is analysed, with a particular emphasis on the Green Deal and the new Ecodesign for Sustainable Products Regulation (ESPR).

The role of industry in this shift towards circular models is described, focusing on the importance of the design phase of products. Ecodesign encompasses comprehensive planning, from raw material extraction to manufacturing technologies, logistics, and end-of-life recyclability. Despite the straightforward concept of creating sustainable products, real-world practices often fall short. Issues such as poor product design and the resultant downcycling, particularly in plastics, hinder the realization of a circular economy.

The author then looks at the role of governments and legislation, identifying the challenge of balancing clear regulatory frameworks with the flexibility for industry innovation as a key one. A hypothetical policy scenario illustrates the complexities and potential of regulating the entire carbon footprint of products, rather than just direct emissions. Effective incentives should combine fiscal, administrative, and reputational elements to stimulate both producers and consumers towards sustainability. Creating markets for sustainable products requires synergistic legislative measures and consumer incentives.

Disclaimer: The information and views set out in this chapter are those of the author and do not necessarily reflect the official opinion of the European Commission.

Michele Galatola, European Commission, Directorate-General for Internal Market, Industry, Entrepreneurship and SMEs (DG GROW)

https://doi.org/10.1515/9783110767179-002

The path to a sustainable economy is multifaceted, demanding substantial changes from industries, governments, and citizens alike. This transition involves long-term investments, procedural reforms, and perhaps most challengingly, shifts in attitudes and behaviors. The journey towards sustainability is arduous, but it is indispensable for securing a viable future.

In essence, this chapter underscores the urgency of transitioning to a circular economy, emphasizing that success hinges on comprehensive collaboration and innovative policy measures. The interplay between environmental imperatives and economic practices forms the crux of Europe's quest for sustainability.

2.1 Introduction

In Europe we are currently facing multiple challenges with interconnected effects that are often difficult to identify and analyze.

Even focusing this analysis only on the environmental dimension of the current challenges, we are faced with various kind of impacts (climate change, biodiversity, water scarcity, land quality deterioration, etc.), different geographical scope (global and local), and different timing (e.g., the impact of air pollution is immediately perceived by the affected population, whilst a CO_2 molecule will exert its climate change impact for the entire time it remains in the atmosphere).

Such a complex situation requires appropriate assessment methods and, from a policy-making perspective, innovative approaches. Life Cycle Assessment (LCA), initially standardized in the ISO 14040:44 set of standards, is the *de minimis* tool required to understand what the relevant environmental impacts are in relation to the life cycle of a product or service, and, in which part of the life cycle they arise.

The European Commission has developed what is currently considered the "best practice" method when it comes to LCA: the product and organization environmental footprint methods (PEF and OEF, respectively). These sets of methods are more operational, detailed, and prescriptive compared to most of the alternative existing LCA-based methods. They have been developed by the Commission and then road-tested on more than 25 different industry sectors (both food and nonfood) in the period 2013–2018. Both the methods and the road-testing have been developed in close collaboration with more than 300 companies, including SMEs, operating in all continents. More than 2,000 stakeholders have actively contributed to the many consultation phases endeavored during this process. PEF and OEF methods [1] are now more and more used in EU policies whenever there is a need for reliable, verifiable, and comparable life cycle data on products and organizations.

Having access to reliable information is the first but ineluctable step in the processes to correctly manage a supply chain. This does not apply only to the final manufacturer who places the product in a market, but is also equally important for all players along the supply chain.

As we will see in a later section of this chapter, developing new tools and business management practices to enhance visibility and transparency along supply chains is one of the success factors that industry needs to consolidate in the short term to have higher chances of succeeding in an economic environment where access to relevant information is key.

2.2 Why products are not already sustainable?

Until recently, energy and natural resources have always been available at relatively low prices worldwide. This has been the main driver against optimization of manufacturing processes and value chains management, as the added value of such changes would not justify the investment and costs needed to achieve them.

The situation has been evolving, building on increased awareness about the environmental damages created by the current economic models, which are based on linear production processes and nearly free rider pollution strategies. The evidence of negative effects on human health and environment about anthropic waste and emissions led to progressively more stringent legislations. This classic "command-and-control" approach led to substantial progress in terms of reduction of pollution but did not provide sufficient drivers to change the underlying economic and production models.

However, notwithstanding the increased amount of legislation and the ambitious environmental targets and objectives we currently have in Europe, the results of these efforts are still insufficient. Depending on how (and which data) are read, the impression ranges from rather positive to quite negative. For example, according to EUROSTAT data related to the first two decades of XXI century, the European gross domestic product (GDP) increased by about 18% whilst the direct material consumption (DMC) decreased, along the same time frame, by about 12%. In abstract, this would be excellent news, as it would indicate an absolute decoupling of economic growth from resource use. However, a more attentive analysis of data would indicate that during this period Europe was hit by a severe economic crisis, with important cuts in production capacities in many relevant (and polluting) sectors.

While DMC is a good indicator for resource use, it ignores the effects on the environment related to the extraction and consumption of those resources. Consequently, it makes sense to look at evolution of GDP compared to changes in the environmental footprint of our economy. In this respect, it is important to distinguish between domestic footprint (the environmental footprint related to the products that are produced in Europe, minus those that are exported) and consumption footprint (that corresponds to the environmental impact of production in Europe plus imports minus exports).

Figure 2.1 illustrates the results of such comparison; these results are included in a study carried out by the Commission Joint Research Centre (JRC) [2].

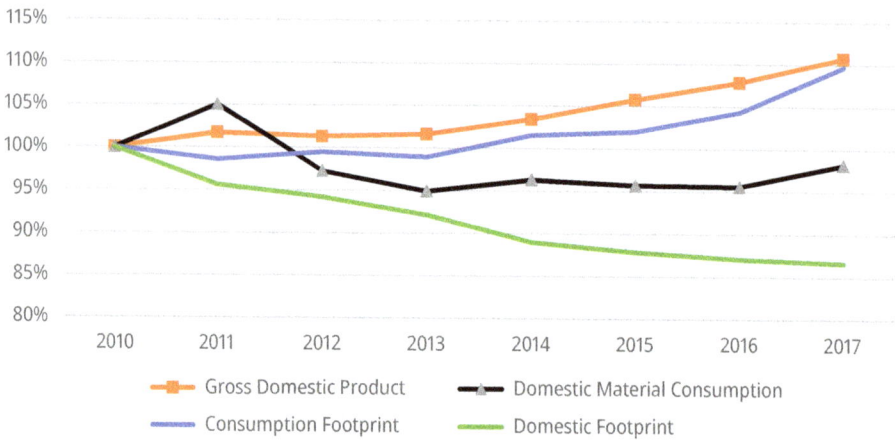

Figure 2.1: Comparison of GDP, DMC, and footprint indicators for the period 2010–2017.

These data clearly illustrate how Europe's results in terms of reduction of environmental impact are mixed. The domestic footprint has decreased substantially in the first part of the twenty-first century, but the environmental impact of our economy has increased, mostly due to the impact of the imported products. For example, while a lot of attention has been paid to reducing the greenhouse gas (GHG) emissions within Europe in an attempt to limit climate change impacts, it cannot be neglected that a large percentage of the carbon footprint of many products stems from indirect emissions (often referred to as scope 3 emissions), that is, emissions not related to the final product itself but rather to activities and processes taking place along the supply chain, often outside Europe.

JRC data clearly show that decoupling is not occurring and we need to become much more attentive, both as consumers and legislators, not only to what we consume, but also to how those products are produced. This change of perspective has important implications from a geopolitical perspective but also in terms of legal design of future policies.

2.3 The role of circular economy

The president of the European Commission, Ursula von der Leyen, emphasized, soon after taking office, in her agenda for Europe that "We need to change the way we produce, consume and trade" [3]. This objective is reflected in the European Green Deal [4], Europe's growth strategy to build a fairer and wealthier society with a cutting-edge, competitive, climate-neutral, and circular economy. The same concept has been further articulated in the European Circular Economy Action Plan [5], adopted by the European Commission in March 2020. The key goal of this action plan and all the initiatives included therein is to influence the internal market to produce and consume more sustainably while easing social and environmental stresses.

Experts seem to agree that critical raw materials (e.g., lithium, tungsten, indium, etc.) that are essential for industry could run out within a few decades if we do not rethink how we use materials in our linear "take-make-dispose" economy. The price volatility and the price of important commodities will continue to increase unless a clear and quick paradigm change is implemented at all scales.

However, while there is clear evidence that shifting to a circular economy makes a lot of business (and sustainability) sense, it is also equally clear that it is not happening, or at least not at the pace needed to reverse a situation of quick and progressive overcoming of our planetary boundaries.

There are multiple reasons why this is happening, or better, not happening. The economic gains related to circular economy are real, but they only materialize when multiple actors across the value chain play their respective role. So, who are these actors and what is the role they should play? First, business players are at the center of any action intended to increase the long-term sustainability of our society; other key actors in this challenge are research communities, investors, governments, and consumers.

Changes in business models are required. In addition to the physical risks of climate change and other repercussions of pollution, businesses must also consider the transition concerns brought on by the shift to net-zero emissions. Businesses will need to restructure and adapt to new legislation and changing consumer preferences as governments begin the decarbonization process. It will not be easy, but the transition should be just and leave no one behind. Some companies will go out of business, and some industries will suffer more than others. It will be necessary to create new academic pathways and gain new abilities.

From the business community perspective, sustainability and circularity have become an area of competitive advantage; companies that fail to adapt run the danger of falling behind more progressive competitors. Moreover, there are direct costs related to the climate crisis and other environmental impacts that are seriously affecting business operations worldwide, requiring the proper adaptation of enterprises in many sectors and geographical areas.

The current historical challenges we are facing, due to the COVID-19 pandemic first and now Russia's unjustified aggression against Ukraine, are forcing the world, and Europe in particular, to rethink and redesign our supply chains and industrial ecosystems. There is an ongoing attempt to bring back strategic manufacturing activities and related supply chains within Europe (but the United States is doing the same) because this is considered strategic from a geopolitical economic as also environmental viewpoint. It helps reducing exposure risks, especially those related to higher commodity prices and risks of supply chain disruptions. More geographically grouped supply chains help address one of the three main barriers to the change required by circular economy, that is, geographic dispersion (the other two being materials complexity and linear lock-in).

2.4 The journey towards an environmentally sustainable economy

There is no magic solution or silver bullet: moving away from the current path and bringing our society towards a sustainable future will require important changes from everyone: industries, governments, citizens. The changes required are of different nature, scale, and implications: some of them require long-term investments, some require change of procedures, and probably the most difficult ones will require a change of attitude. Without wanting to step into the area of human psychology, not falling under this context, it is still evident that modifying habits and *status quo* is one of the most difficult endeavors as there is a human tendency to remain within the boundaries of what we are used to. This is especially true when the drivers of change are not immediately perceivable by the actor who is asked to make that change.

In the following sections the role of some main stakeholders' groups will be discussed to highlight role and responsibilities of each one.

2.4.1 The role of industry

The way products are "designed" determines up to 80% of the life cycle environmental impacts they cause [6]. It is interesting to note how the understanding of the term "design," or eco-design in this context, is vastly different from the industry perspective compared to, for example, the legislator's one.

From a manufacturing perspective, eco-design is a holistic concept that includes all the activities that directly or indirectly determine the environmental characteristics of a product. Therefore, the implications of where raw materials are extracted, what manufacturing technologies will be used, the choice of materials, the logistics, the after-sales services provided, the operations needed to recycle the product (in case an extended producer responsibility scheme is in place), are all components of the eco-design of a product. This holistic definition of what eco-design means offers a lot of flexibility to industry actors; an improvement of the eco-design of a product would not necessarily require an active change in the materials or manufacturing process, but could be equally achieved through a rationalization of the logistics or a change in some of the suppliers involved. Leaving this flexibility to the industry operators, they would be able to identify the solutions that optimize the cost/benefits compared to objective to be achieved.

From an industry perspective, therefore, to manufacture an "environmentally sustainable" product means to carefully design every single step in the value chain to: (i) extend the lifetime of the product as much as possible, (ii) minimize the use of natural resources and energy, (iii) minimize the release of pollutants in the air, water, and soil, and (iv) maximize the recyclability of the product at the end of its functional life. Eventually, it could be said that an environmentally sustainable product is one that minimizes its life cycle environmental footprint.

However, while the "recipe" for a sustainable product seems quite straightforward, the analysis of what happens in the economy seems to show a different reality. For example, one of the most significant contributors to ongoing linearity and reliance on virgin resources, according to the 2020 Circularity Gap report, is poor product design [7]. The analysis of the circularity features of a product (i.e., durability, reparability, recyclability, reusability, etc.) reveals some tendencies going rather towards a reduction of products' circularity. There is evidence that the lifespan of some products is reducing [8–10]; whether this is intentional or not it is the subject of a separate and difficult conversation to be held on "planned obsolescence" (but that it is outside of the scope of this chapter). This tendency is sometime matched with a product design that makes dismantlement and reparability of some product components difficult, if not impossible (e.g., the impossibility to replace the batteries in some models of smartphones).

Similar tendencies are shown when it comes to recyclability. This is particularly evident in the case of plastic polymers. The purity of a material is often the limiting factor for its recyclability; if a recycler is faced with flows of different materials with each material being blended with all sorts of additives, then the possibility of having a "closed-loop" recycling process, that is, a process producing a recycled material having the same properties as the virgin one, is minimized. This leads to the unavoidable downcycling of plastic polymers, a by-effect of product design with very negative environmental and economic implications [11]. This is not meant to imply that additives are not useful or that different polymers should never be used in the same product. Though there might be good reasons to use additives, this is not the only (and not necessarily the best) way and here is where a cultural evolution is needed, either pushed by customers' pressure or induced by wise legislation.

2.4.2 The role of governments and legislations

The approach to the eco-design concept from the legislator's perspective is different compared to the industry's perception. In the vast majority of the eco-design legislations currently in place, in Europe and worldwide, the notion of eco-design has been implemented in a rather restrictive way, focusing on environmental characteristics or performance shown by a product "as placed on the market." In many cases, this means that any legal requirement related to a product should refer to properties that can be measured/tested through laboratory tests, often based on European/international standards.

This approach has clear advantages in terms of implementability and enforceability of a policy. It provides a relatively easy framework to identify key relevant environmental characteristics of a product and regulate them with the intention of achieving the specific objectives of a certain policy. From the companies' perspective, this approach gives them a clear set of rules they must comply with, reducing complexity and administrative burdens (with the related costs).

However, this approach also brings with it several important limitations and drawbacks, both for companies and regulators (and at the end of the day, for citizens who should benefit from the corresponding legislations).

A narrow interpretation of the eco-design concept highly reduces the freedom of industry operators to identify the solutions that would be more convenient to them, from an organizational and economic perspective. By focusing the legal requirements only on properties/characteristics that can be measured or tested on a product, the legislator somehow imposes certain eco-design choices, which are related to materials, chemicals, or manufacturing processes to be used. This prescriptive approach could potentially lead to sub-optimal solutions, limit product-related innovation, and may even be not completely effective to achieve the intended environmental objectives.

Let us consider a purely hypothetical scenario: in the context of the Paris Agreement a government decides to implement a national policy to reduce their national carbon footprint, making important economic incentives available to national manufacturing companies who cut their direct GHG emissions (i.e., the emissions for which they have a direct operation control). The rationale of this government is solid: they are operating in a territory on which they have jurisdiction, and they are targeting emissions that, though not being an intrinsic property of the products sold, can be audited through standards and consolidated procedures. Therefore, from an implementability and enforceability perspective, the design of this policy is not questionable. However, the effectiveness of such a policy would be questionable for the following reasons:

1. Climate change is a global impact: it does not matter if a molecule of CO_2 is emitted in Europe, Asia, or America, it will have the same negative impacts on all citizens of the world. This notion could be even extended to the point of affirming that it is the responsibility of each government to introduce legislations protecting the health of their citizens from environmental impacts affecting them, independently from where those impacts are generated.
2. The quantity of emissions targeted through such policy would be mostly a minority of the total carbon footprint for each product placed on the market of that country. This is even more the case as the policy would only target products produced within the territory regulated by that government.
3. By limiting the measure to only direct emissions caused by the manufacturing processes, this policy would hinder the possibility for businesses to identify better

solutions to achieve the same policy objective (e.g., change of materials, use of renewable energy sources, change of suppliers, change of logistics, etc.).
4. By only focusing on one environmental impact, even as relevant and urgent to be tackled as climate change, there is the risk that other potentially relevant impacts may excessively increase.

An alternative approach this government may take would be to identify a threshold, maybe in line with the climate change planetary boundary, for categories of products sold in their country (the shift to a consumption-based approach would be essential). Once that threshold is identified (e.g., x kg CO_{2eq} for product group A, y kg CO_{2eq} for product group B, z kg CO_{2eq} for product group C, etc.) each manufacturer would autonomously identify the best strategy to "eco-design" products that can lie within the identified threshold.

This approach, however, would also introduce procedural complexities, probably an increase of costs for companies (at least in the short term), and it would require different set of tools and skills, especially on the government side. For example, the introduction of a requirement not related to a characteristic measurable or testable on a product but rather something intangible and related to the product entire life cycle (i.e., the product carbon footprint) would also require the following complementing measures:
–   Development of a set of harmonized rules to calculate the carbon footprint for each product group regulated. The rules should be the same for any economic operator, as this is a prerequisite to enhance reproducibility, reliability, verifiability, and comparability of the information produced.
–   Availability of common secondary life cycle datasets to be used in the calculation of the carbon footprint for all activities where a company does not have operational control or access to primary data.
–   Availability of a simplified IT tool to be used by companies for free to calculate the carbon footprint of their products in compliance with the harmonized rules (this feature would be optional, but it would provide benefits in terms of costs reduction for companies and implementability of the legislation, especially by SMEs).
–   A system to convey and make available this information digitally, simplifying the secured flow of relevant information along the supply chain.
–   A wide network of accredited third-party verifiers able to certify the calculations carried out by companies to have reasonable assurance about the reliability of the data produced.

A last note related to this hypothetical example is on the incentives that this government could make available. Experiences in various regions of the world have shown that the most effective environmental incentives to companies include a wise mix of fiscal, administrative, and reputational incentives. However, these may not be sufficient if not combined with incentives that could also induce consumers to change their shopping behaviors. In other words, it is not sufficient to increase the availability of sustainable products if there is no market ready to absorb them.

The creation of these "lead markets for sustainable products" is something that a government could aim at by working synergistically at the level of inducive legislations (like the one described above), providing financial benefits to companies producing such products (e.g., fiscal incentives), creating dedicated markets for such products (e.g., via green public procurement). Moreover, a decisive and innovative choice would be to offer direct economic benefits to consumers to stimulate the choice of more sustainable products.

For example, the policy described for this hypothetical government, could also introduce an economic incentive to consumers to buy products with a low carbon footprint. Each citizen would receive an annual "carbon budget" corresponding to the regulated products. Each product sold would show a certain number of points, proportionate to the difference between the average carbon footprint for that product group and the specific carbon footprint of the product bought. This information would be transferred to the personal digital wallet of that citizen, in full compliance with privacy rules and without allowing commercial exploitation of that information. At the end of the year, the addition of all the points saved (or lost, in case the citizen buys a product with a worse carbon footprint than the average), would be converted by the government into a "premium" (cash, tax reduction, green check, free public transports, etc.). This system would establish a clear and direct link between an action performed by the citizen (buying something) with a consequence of that action (saving carbon budget or not), promoting virtuous habits (buying more sustainable products or not buying them at all).

2.4.3 The role of citizens as consumers

Too often, in the author's opinion, citizens in their role of consumers are held responsible for "doing the right thing": they are expected to buy fewer products, buy more sustainable products, recycle more, etc. While there is undoubtedly a pivotal role to be played by each of us in our private life, while it is true that "every single drop counts," a "consumer" is often asked to solve very complex puzzles within a very limited time span.

There is an "information dilemma" related to sustainability of products: studies have proven that too much information is often equivalent to no information; on the other hand, there are also studies showing that consumers claim that the lack of sufficient relevant information and affordable alternatives prevents them from choosing sustainable products. About 85% of consumers are dissatisfied with the information that is currently available to them, as it is not helping them to decide whether to switch to buying more environmentally friendly items [12, 13].

The key determinant is probably related to the "relevancy" of the information. On one hand, it is difficult to identify information that is horizontally relevant for all consumers: some people would consider climate change as the environmental determinant factor in driving their choices, and therefore could be interested in carbon footprint scores for each product, while others could care more for ecosystem health and therefore rather look at biodiversity and its proxy indicators as a better compass to guide their choices. On the other hand, there are limiting factors to the amount of information that can be provided relative to a product. Nowadays, most of the information that the consumers look for is available on packaging or on labels attached to a packaging. This is an effective way to provide key information, but there are physical limitations to what can be put on a label or packaging while keeping it readable. However, this may soon be a problem of the past, thanks to new technologies and legislative measures currently under development. Digitalization of product-related information is making more and more information directly available online, usually through scanning of data carriers (e.g., barcodes or QR codes) through our own smartphones.

The European Commission has recently introduced in the revised Eco-design for Sustainable Product Regulation (see Section 2.5), a new tool called digital product passport (DPP). The objective is to make available in digital form all the relevant life cycle-, environmental, and circularity-related information for products regulated through the eco-design framework. The DPP will serve multiple purposes:

- It will allow companies to rationalize the way they make available information along their value chain, both to their economic counterparts but also to fulfil legal requirements. It will also increase the business capacity to track and trace information, an essential feature to limit disruptions of supply chains like those experienced recently due to the COVID-19 pandemic and Russia's unjustified aggression against Ukraine.
- It will facilitate compliance controls by customs and national market surveillance authorities.
- It will provide governments a powerful tool to monitor the environmental and circularity performance of products placed on the European market (including the imported ones).
- It will allow citizens to have access to the environmental and circularity-related information that they deem relevant before making their own purchases.

When it comes to the potential use of DPP for consumers, it is expected that dedicated applications will be developed to filter the information available in the passport and provide to the consumers the info they are interested in. This is something that is already happening for food or cosmetics. There are, for example, apps developed on a voluntary basis that provide information on the content of some specific chemicals or nutrients, provided that the information is available digitally. This is possible through scanning a data carrier placed on the product or its packaging and the use of dedicated application programming interfaces (APIs), often available for free in the most common apps web shops.

Another element of high importance is related to the affordability of sustainable products. The COVID-19 pandemic has greatly increased public awareness of environmental issues. Just to give an example, according to 55% of US customers [14], the pandemic has increased their desire to purchase environmentally sustainable products. However, in many cases the price of sustainable products is much higher than that of conventional products (i.e., products that do not bear environmental labels or have green claims attached to them). Market analyses show that "sustainable products" are sold at higher prices than grey products, with differences varying widely depending on the category the products belong to. On average, it seems that beauty and health products are those that show the widest markup compared to equivalent grey products, up to 200%. Differences are less relevant, but still important for food products and energy. However, only a minor part of these higher costs can be accounted for the different "design" used to produce sustainable products. According to Kearney study [14], only 10–30% of the final cost of a product is due to its production process; all the other costs are related to the branding, downstream costs, and profit margins.

Research activities, use of low-impact materials, low-pollution technologies, dedicated human resources devoted to sustainability-related administrative and management tasks, all tend to increase the manufacturing costs. However, a sustainable manufacturing process also leads to economic savings related to lower energy consumption, lower waste production, more efficient processes, better access to finance, lower risks of environmental accidents (and therefore lower insurance premiums), etc. The additional costs due to the sustainability features of the products are estimated at around 10% [14]. And this is close to what consumers would be willing to pay more for a sustainable product. The same Kearney study shows that "around 70% of all consumers would pay up to 10% more, another 15% would pay 30% more, and another 15% will pay even higher markups."

While there are no conclusive scientific studies showing where the additional price gap stems from, anecdotal evidence indicates that companies artificially increase the price of sustainable products because "that is what consumers expect": it seems that retailers have own evidence that consumers would become suspicious of a sustainable product marketed as green that would not cost more than a conventional one. This information bias therefore causes (in part) the increased price gap between conventional and sustainable products and, at the same time, limits the further uptake of the sustainable ones.

2.5 The new proposal for a regulation on eco-design of sustainable products (ESPR)

At the core of the 2020 Circular Economy Action Plan there is a new framework regulation on the eco-design of sustainable products. This proposal for a regulation has been adopted by the Commission[1] in March 2022. It is an extensive review of an already existing legislation, the Eco-design Directive.[2]

The ESPR framework extends the scope of the existing Eco-design Directive along three main axes:
1. It expands the number of products in scope, moving from the current scope ("energy-using" and "energy-related" products) to all products except for food, feed, and some pharmaceuticals.
2. It widens the scope in terms of environmental aspects to be potentially regulated. While the existing Eco-design Directive was mainly regulating energy-related parameters and impacts, with some valuable addition of some circularity-oriented parameters for some product groups like washing machines and dishwashers, the new ESPR will focus much more on environmental sustainability at large (i.e., including but going beyond energy) and circularity of products. Aspects like carbon and environmental footprint, recycled content, recyclability, and biodiversity will be systematically analyzed and, whenever relevant and appropriate, regulated at product group level.
3. It aims at looking more systematically and consistently also at parameters that are not specific to the product as placed on the market. This possibility, already possible within the current legal framework, has never been used till now. It refers to the possibility to introduce mandatory requirements indirectly related to how a product is produced. The objective is to remain technology-neutral (not prescribing manufacturers how to run their production process), but at the same time to regulate those environmental aspects that have a relevant life cycle sustainability and/ or circularity impact (e.g., the life cycle carbon footprint related to the product).

The ESPR regulation proposed by the Commission introduces also other important elements aimed at increasing the sustainability and circularity of the products placed on the European market, like the DPP, mandatory green public procurement criteria also based on environmental performance classes, and the potential ban on the destruction of unsold goods.

---

1 At the time of drafting this chapter [October 2022], the proposal is in co-decision, meaning it is discussed with the European Council and the European Parliament with the objective of agreeing on a shared final text.
2 Directive 2009/125/EC of the European Parliament and of the Council of 21 October 2009 establishing a framework for the setting of eco-design requirements for energy-related products.

The ESPR proposal could be a game changer in terms of how product-related policies are designed and it is expected to bring important positive impacts also beyond Europe. It strengthens an already existing and successful policy framework, and it introduces powerful bridges to bring coherence and consistence among different pieces of legislation. For example, the deployment of the digital product passport will allow stakeholders to have digital access to product-related information relevant to them through a single-entry point. This will enable great potential benefits to businesses (e.g., in terms of supply chain management and reduction of administrative burdens/ costs related to production of data required by existing laws), to market surveillance and customs authorities (making their checks more efficient and effective), to governments (making available relevant data related to the environmental and circularity performance of the products placed on the European market), and to citizens (allowing them to have access to information relevant for their shopping decisions). It will also support the creation of new business opportunities, especially in the area of re-manufacturing, recyclability and reparability. Last, but not least, it could enable the creation of lead markets for sustainable and circular products, simplifying the link between the environmental life cycle performance of a product and potential incentives at EU and national level.

2.6 Conclusions

The limits of linear consumption have been reached. On both a micro- and macroeconomic level, a circular economy has advantages that are tactical and strategic. There is tremendous potential for innovation, economic growth, and job creation. Businesses are beginning to understand that this linear structure exposes them to increased risks, such as rising resource prices and supply disruptions. Moreover, the recent international crisis has clearly shown that long, complexly integrated global supply chains raise the risks to supply security and safety.

Circular economy and legislations promoting it, in combination with environmental sustainability of products and services, are important contributors to reverse the current situation and create the conditions for a new and vigorous paradigm change in our economies and society. National governments and the European Commission are working together to enable the changes required.

Since the publication of the first EU Circular Economy Action Plan in 2015, the European Commission has embraced circularity as the new economic paradigm for Europe.

New legislation, such as the recently proposed Eco-design for Sustainable Products Regulation (ESPR), are helping to offer a systemic solution to decreasing reliance on resource markets, reducing susceptibility to resource price shocks, and gradually internalizing the societal and environmental "externality" costs that businesses do not bear. But no legislation will ever be truly effective without the active and coordinate contribution of citizens, civil society organizations, and industry associations.

References

[1] The European Commission product environmental footprint (PEF) and organisation environmental footprint (OEF) methods are downloadable at: https://environment.ec.europa.eu/publications/rec ommendation-use-environmental-footprint-methods_en

[2] Sanyé-Mengual, E., Secchi, M., Corrado, S., Beylot, A. and Sala, S., Assessing the decoupling of economic growth from environmental impacts in the European Union: A consumption-based approach. Journal of Cleaner Production, 2019, 236, 117535.

[3] Von der Leyen, U., – A union that strives for more – my agenda for Europe. Page 7. 2019, Available at: https://ec.europa.eu/commission/sites/beta-political/files/political-guidelines-next-commission_en.pdf

[4] Communication from the Commission to the European Parliament, the Council, the European Economic and Social Committee and the Committee o Regions: A new Circular Economy Action Plan For a cleaner and more competitive Europe – COM/2019/640 final

[5] Communication from the Commission to the European Parliament, the Council, the European Economic and Social Committee and the Committee o Regions: A new Circular Economy Action Plan For a cleaner and more competitive Europe – COM/2020/98 final.

[6] "How to Do Eco-design?", a Guide for Environmentally and Economically Sound Design Edited by the German Federal Environmental Agency. Verlag form, 2000.

[7] Circularity Gap Report. 2020, 15. https://assets.website-files.com/5e185aa4d27bcf348400ed82/ 5e26ead616b6d1d157ff4293_20200120%20-%20CGR%20Global%20-%20Report%20web%20single% 20page%20-%20210x297mm%20-%20compressed.pdf

[8] EEB. Cool products don't cost the earth -full report. 2019, www.eeb.org/coolproducts-report

[9] "Halte à l'obsolescence programmée – HOP." "Lave-linge: Une durabilité qui prend l'eau?." 2019, https://www.halteobsolescence.org/wp-content/uploads/2019/09/Rapport-lave-linge.pdf

[10] Geplante obsoleszenz: Entstehungsursachen, konkrete beispiele, schadensfolgen, handlungsprogramm – gutachten im auftrag der bundestagsfraktion bündnis 90/die grünen. 2013, https://www.schridde.org/download/Studie-Obsoleszenz-aktualisiert.pdf

[11] Plastics recyclers Europe. https://www.plasticsrecyclers.eu/challenges-and-opportunities

[12] Commission staff working document: Sustainable products in a circular economy – towards an EU product policy framework contribution to the circular economy. 2019, 92 final.

[13] Nicolli, F., Johnstone, N. and Söderholm, P., Resolving failures in recycling markets: The role of technological innovation. Environmental Economics and Policy Studies, 2012, 14, 261–288.

[14] Kearney's Earth Day 2020 consumer sentiments study. accessed October 17 at https://www.nl.kear ney.com/consumer-retail/article/-/insights/why-todays-pricing-is-sabotaging-sustainability

Pengxian Ye, Debjyoti Ray, and Chunshan Song[*]

3 Circular economy for CO_2 utilization and hydrogen production in China

Abstract: The mounting environmental, economic, and societal challenges in the world led to a shift of thinking from the linear economy to the circular economy (CE). With the development of CE in recent decades, the scope of CE in China is no longer limited to waste management, but covers a broader range in line with tackling climate change, where CO_2 utilization as a resource plays a key role. Hydrogen energy development is of vital importance in reducing the greenhouse gas emissions and establishing a low-carbon society. Hence, the latest policies on CE, CO_2 utilization, and hydrogen production in China are summarized in this chapter. Moreover, the major technologies for CO_2 utilization and hydrogen production in China are introduced with relevant cost analysis. Both the policy support and the advancement of technologies are needed to tackle the current challenges for further development of CO_2 utilization and hydrogen production in China.

3.1 Circular economy in China

3.1.1 The development of circular economy in China

Since the late 1970s, the concept of circular economy (CE) has been gaining increasing attention in the world due to the mounting environmental, economic, and societal challenges [1, 2]. As shown in Figure 3.1, different from the traditional linear economy model following the take-make-dispose pattern, CE is to establish a closed-loop system based on the 3R principle (reduce, reuse, and recycle), aiming to balance economic development with environmental protection by maximizing the use of renewable resources, recirculating resources and products, and also designing waste out of the economic system [3–6].

The concept of CE in China had origins in the 1990s on cleaner production, industrial ecology, as well as ecological modernization [8]. The development of CE in China was inspired by that in Europe, the United States, and Japan [9, 10]. With the development and implementation of CE in China, the main focus of the CE shifted gradually from narrow waste recycling to broad efficiency-oriented control during the closed-

[*]**Corresponding author: Chunshan Song**, Department of Chemistry, Faculty of Science, The Chinese University of Hong Kong, Shatin, Hong Kong SAR, China
Pengxian Ye, Debjyoti Ray, Department of Chemistry, Faculty of Science, The Chinese University of Hong Kong, Shatin, Hong Kong SAR, China

https://doi.org/10.1515/9783110767179-003

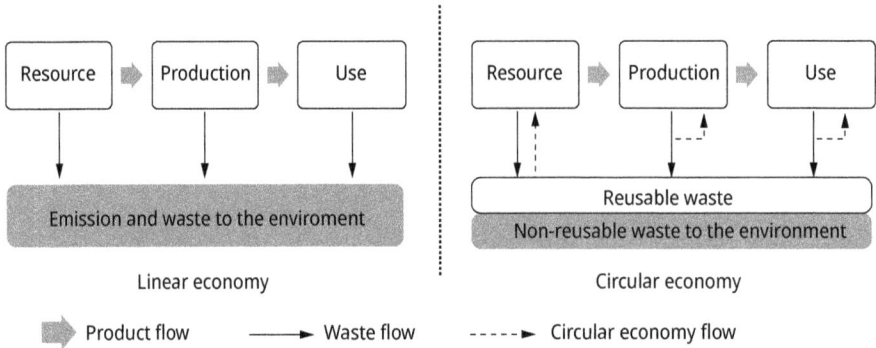

Figure 3.1: Linear economy and circular economy model [7].

loop flows of materials at all stages of production, distribution, and consumption [8]. In 2002, the concept of CE was formally accepted as a development strategy by the government [10]. Over the past 20 years, China has developed wide-ranging ambitions for the CE and enacted comprehensive CE policies and regulations with a top-down approach (Table 3.1) [11]. Based on the Circular Economy Promotion Law, which came into force in 2009, several action plans have been issued for various specific sectors to support the CE framework. Different from some developed countries that mainly focus on CE in waste management and resource utilization efficiency [10], China encounters the more complex environmental problems due to heavy reliance on coal in the primary energy and the rapid and intensive industrialization and urbanization process [12], and thus has great concerns in not only waste disposal but also pollution control and tackling climate change.

Table 3.1: Policies issued for circular economy during 2005–2021 in China [5, 13].

Time	Policy	Main content
2005	Opinions on Accelerating Growth of Circular Economy	Accelerated the efforts to establish and improve the laws and regulations system on circular economy in line with China's national conditions
2006	The Outline of the 11th Five-Year Plan for National Economic and Social Development of the People's Republic of China	Presented a plan for circular economy development, which was included as one of the major strategic tasks in the 11th Five-Year Plan period
2010	Guidelines for Making Plans for Circular Economy Development	Pointed out that local governments should develop the circular economy according to their specific circumstances

Table 3.1 (continued)

Time	Policy	Main content
2013	Development Strategy and Immediate Action Plan of Circular Economy	Set goals for China's circular economy development in different stages to establish a circular agricultural system, a circular industrial system, and a circular service industry system to facilitate the industrialization of resource recycling
2016	The 13th Five-Year Plan for Economic and Social Development	Promoted economic and intensive resource use, responded to the global climate change, and developed green and environmentally friendly industries
2021	Development Plan for the Circular Economy in the 14th Five-Year Plan Period	Developed the circular economy through various initiatives, such as promoting recycling, remanufacturing, green product design, and renewable resources

Even though China has made prominent progress in promoting the development of CE, there are still some limitations currently, including low resource productivity of key industries, low standardization level of renewable resource recovery, lack of land for recycling facilities, high difficulty in the utilization of low-value recyclables, and low utilization rate of bulk solid waste (from "14th Five-Year" Circular Economy Development Plan). In addition to the limitations of pertinent technologies [14], relevant policies in China also remain to be improved. Most policies have not explicitly taken a life cycle perspective, lacking a systematic way to promote the CE [7]. Moreover, market-based policies are still lacking in China. The strong dependence on direct subsidies, tax relief, and financing support policies for national companies may lead to market distortions and cause the economic and social development to deviate from an optimal scale of production and service provision. Considering the current situation of the CE development in China, the government unveiled a new development plan in 2021 to spur the CE in the next 5 years.

3.1.2 The latest policy on the circular economy in China

In July 2021, the National Development and Reform Commission of China released the Development Plan for the Circular Economy in the 14th Five-Year Plan Period (2021–2025) [15]. This plan aims to develop the CE through various initiatives, such as promoting recycling, remanufacturing, green product design, and renewable resources [11]. Moreover, this plan sets several specific targets in resource productivity, energy consumption, and water consumption, utilization rates for different solid wastes, and resource recycling for China to reach by 2025. Besides these development targets, this plan lists three key tasks for the regional governments to accomplish according to their local conditions over the course of the period, including building a resource recycling industry system and improving resource utilization efficiency, building a recycling system for waste materials and fostering a recycling-oriented society, and further development of the agricultural CE and establishing circular agricultural production. Furthermore, this plan puts forward several sector-specific actions such as plastic pollution control, promoting green packaging for shipping and logistics, recycling used batteries, and managing the life cycle of vehicles.

Apparently, this recent plan attaches vital importance to reducing energy consumption as well as increasing resource utilization efficiency and aims to establish an all-round CE framework based on the current situations of China. On the one hand, the CE framework in China helps to lower energy-intensive raw material inputs, increase the efficiency of industrial processes, and improve the recycling of resources [7, 16]. On the other hand, China can reduce the dependence of its economic development on primary resources and alleviate the resource-constraint issues it may face in the process of achieving carbon neutrality with the help of this CE framework [17]. According to the estimation by CACE, developing CE contributes to about 25% of the carbon emission reduction in China from 2016 to 2020 [17]. Hence, the CE forms an important component of China's broader climate strategy to tackle climate change and achieve carbon neutrality by 2060. There have been some literature reviews on CE in China [8, 14, 18], but most of them focus on cleaner production and waste recycling. This chapter provides a summary and brief analysis of the CE policies and relevant technologies in tandem with the latest carbon neutrality goals in China, with main focuses on CO_2 utilization and H_2 production.

3.2 CO_2 utilization in China

3.2.1 Carbon neutrality goals of China

According to the National Oceanic and Atmospheric Administration, the global atmospheric CO_2 concentration rose to about 419 ppm in December 2022 (Figure 3.2) [19]. CO_2 is a colorless and odorless gas playing an important role in Earth's carbon cycle [20]. The asymmetric stretching mode and bending mode of CO_2 vibration create dipoles and enable CO_2 molecule to absorb and reemit the infrared radiation, making the temperature of our earth ramp up. As one of the greenhouse gases, CO_2 is mainly generated from combustion of fossil fuels (coal, petroleum, and natural gas) and dumped into the atmosphere as a waste. Global CO_2 emissions from energy combustion and industrial processes reached 36.3 gigatons in 2021 [21], estimated by the International Energy Agency (IEA). Climate change issues such as global warming or extreme weather events are therefore increasingly grim. These human-induced climate changes are causing dangerous and widespread disruption in nature and affecting the people in the world [22, 23].

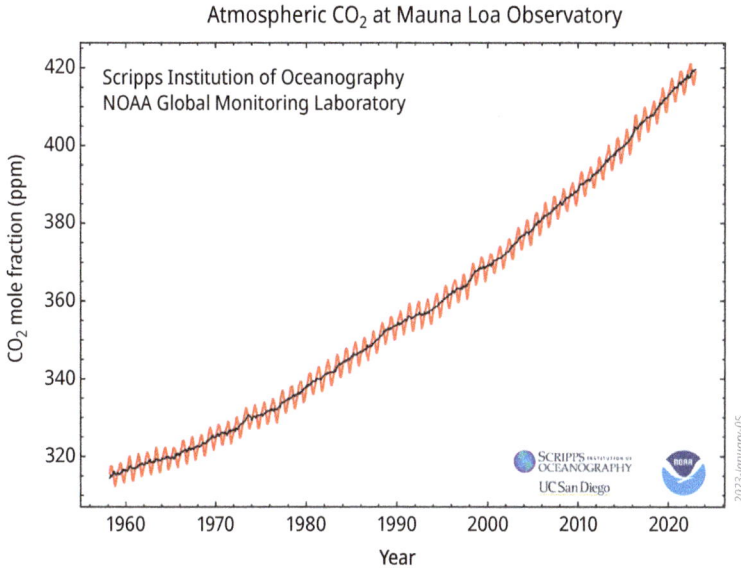

Figure 3.2: Global atmospheric CO_2 since 1958 from National Oceanic and Atmospheric Administration (NOAA). The red curve indicates the monthly mean values and the black one represents the seasonally corrected data [19].

As the largest emitter of CO_2 in the world, China accounts for approximately 30% of global emissions [24]. According to the Carbon Emission Accounts & Datasets (CEADs), CO_2 emissions in China are mainly from power sector (29.91%) and industrial sector (24.91%), followed by residential sector and transportation sector [25]. In 2020, China announced the target of hitting the peak of carbon emission before 2030 and reaching carbon neutrality by 2060. Carbon neutrality refers to net-zero CO_2 emissions, a balance between all anthropogenic emissions and CO_2 removals from land, ocean, and human society [26]. The main objectives for carbon peaking and carbon neutrality were divided into three stages: By 2025, China will have created an initial framework for a green, low-carbon, and CE and greatly improved the energy efficiency of key industries. By 2030, China will see significant accomplishments from the comprehensive green transformation in economic and social development, with energy efficiency in key energy-consuming industries reaching advanced international levels. By 2060, China will have fully established a green, low-carbon, and CE and a clean, low-carbon, safe, and efficient energy system. Energy efficiency will be at the advanced level by international standard, and the share of nonfossil energy consumption will be over 80%. China will be carbon-neutral, and it will have reached a new level of harmony between humanity and nature. Several numerical targets in energy consumption, CO_2 emissions, the share of nonfossil energy consumption, the forest coverage rate, and stock volume were set clearly for each stage, which will lay a solid foundation for carbon dioxide peaking and carbon neutrality.

To achieve the carbon peaking and carbon neutrality goals, China has formulated a "1 + N" policy framework, where "1" means the long-term approach to combating climate change documented in Working Guidance for Carbon Dioxide Peaking and Carbon Neutrality in Full and Faithful Implementation of the New Development Philosophy [27], and "N" refers to several solutions to achieve peak carbon emissions by 2030 such as the Action plan for carbon dioxide peaking before 2030 issued in October 2021 [28], and the Sci-tech action plan to support carbon peaking and carbon neutrality (2022–2030) issued in August 2022 [29]. This policy framework sets several tasks such as strictly controlling fossil fuel consumption, actively developing nonfossil energy, deepening reforms of energy systems and mechanisms, as well as accelerating the development of a clean, low-carbon, safe, and efficient energy system, and supports these tasks with a series of recent policies and regulations. Besides, several tailored action plans were formulated to advance actions on carbon dioxide peaking and carbon neutrality in China. Among these plans, promoting the development of renewable energy, energy storage, hydrogen energy, and carbon dioxide capture, utilization, and storage (CCUS) is of great significance.

To achieve carbon neutrality, two directions need to be focused on. One is to reduce CO_2 emissions directly through measures like changing energy portfolio, economic structural optimization, energy conservation, and technological advances [30]. Coal reduction and gradual phasing-out as well as increasing renewable energy are crucial for China to achieve carbon neutrality by 2060 [26]. The other direction is to use negative emission technology to offset the emissions, where CCUS will play an important part. According to different net-zero emission pathways studied by Chinese research institutions, CCUS technology is essential to attaining the carbon neutrality goal, with a contribution to emission reductions of 0.6–1.6 billion tons in 2050 [31, 32]. Moreover, CO_2 utilization is in line with the concept of CE since it regards CO_2 as a resource rather than a waste, and then recycles or reuses it to close the carbon cycle. Contributions needed to achieve carbon neutrality in China are summarized in Figure 3.3 [26]. Specifically, negative emission actions, for instance, CCUS technology, are expected to contribute nearly 25% of total emissions reduction from 2020 to 2060. In this section, the focus will be on the current status of CO_2 utilization in China.

3.2.2 Status of CO_2 utilization technology in China

CCUS is a promising part of the strategy for reducing the atmospheric concentration of CO_2 to tackle the climate change issues and achieve circular carbon economy [33]. In particular, CO_2 utilization, which is defined as the process of using CO_2 as a raw material for products or services with a potential market value [34], satisfies the 3R

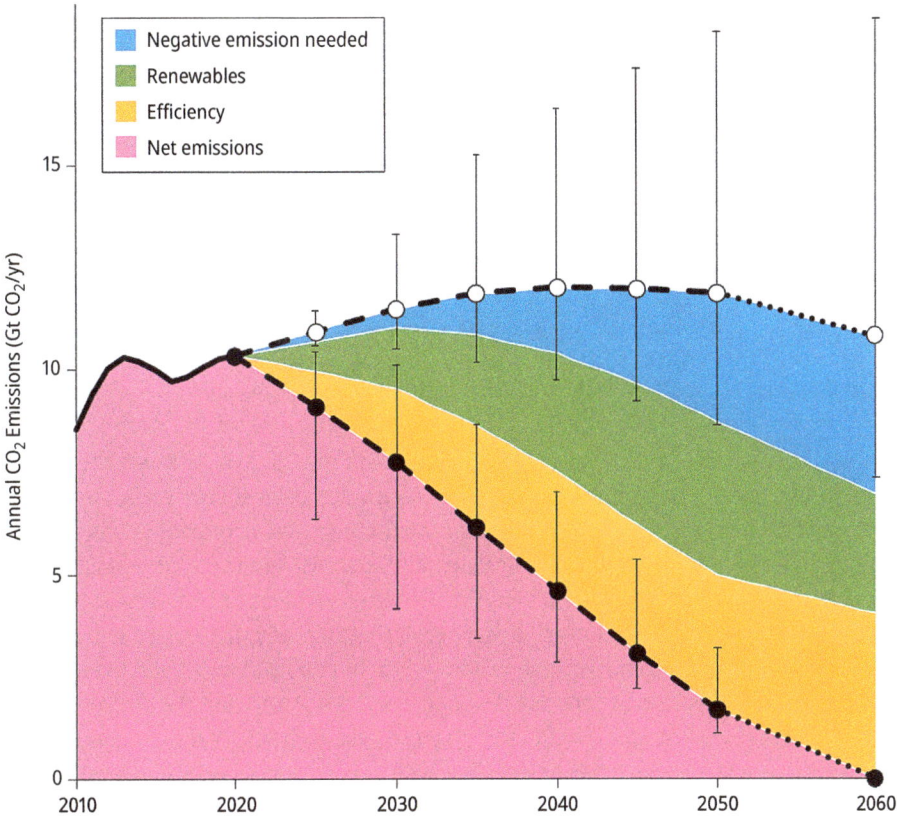

Figure 3.3: Contributions needed to achieve 2060 carbon neutrality target in China [26]. This is the illustration of how negative emission actions, renewables, and improved efficiencies can reduce net emissions, and thus achieve carbon neutrality by 2060 in China. Open circles denote the average baseline scenario emissions from five models, and the solid black circles indicate the average emissions from the same models but with the 1.5 °C limit. The error bars show the max–min spread of these models. "Efficiency" denotes end-use efficiency and fossil fuel subsidies reform, and "negative emission needed" covers nuclear, CCUS, and other technologies.

principle of the CE as it contributes to reducing CO_2 emissions and creating a closed carbon cycle by CO_2 reuse. CO_2 utilization can be divided into direct utilization, where CO_2 is not chemically altered (nonconversion), and the chemical and biological conversion of CO_2 to industrially useful products (such as chemicals, fuels, and materials) [20, 34–36]. The common CO_2 utilization pathways are summarized in Figure 3.4 [34]. Although still under development, the chemical and biological utilization of CO_2 has drawn increasing attention in recent years. According to the IEA, the amount of CO_2 used in the world each year is around 230 million tons (Mt), among which 130 Mt of CO_2 are used to produce urea, and around 80 Mt are used for enhanced oil recovery (EOR) [21].

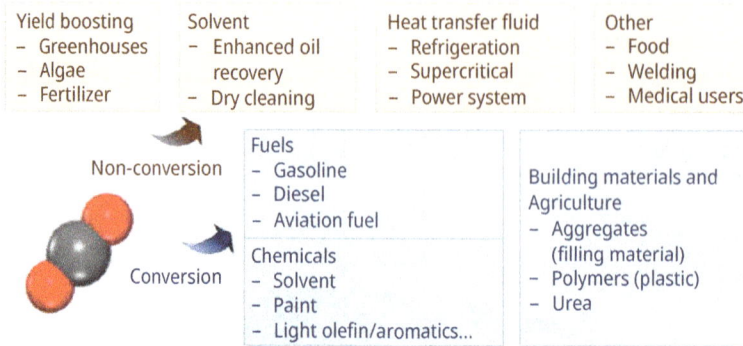

Yield boosting	Solvent	Heat transfer fluid	Other
– Greenhouses	– Enhanced oil	– Refrigeration	– Food
– Algae	recovery	– Supercritical	– Welding
– Fertilizer	– Dry cleaning	– Power system	– Medical users

Non-conversion

Fuels
- Gasoline
- Diesel
- Aviation fuel

Building materials and Agriculture
- Aggregates (filling material)
- Polymers (plastic)
- Urea

Conversion

Chemicals
- Solvent
- Paint
- Light olefin/aromatics...

Figure 3.4: Simple classification of CO_2 utilization pathways [34].

China accounts for 21% of the current global demand for CO_2 [34]. Through the chemical and biological pathways, CO_2 is utilized to produce value-added fuels, chemicals, and building materials in China. Today, approximately 100 kt/year of CO_2 is used in China to synthesize high-value chemicals with an output value of approximately USD 58 (CNY 400) million/year, while 50 kt/year is used in synthesizing materials, generating USD 29 (CNY 200) million/year in revenues [37]. Currently, the largest scale of CO_2 bioutilization project by microalgae in China reaches 200,000 tons/year, the amount of CO_2 utilized for curing of concrete is about 10,000 tons/year, and the demonstration project of CO_2 mineralization with steel slag in Sichuan province utilizes 50,000 tons of CO_2 per year [32].

Different from chemical or biological utilization of CO_2, geological utilization directly recycles or reuses CO_2 due to its intrinsic physical or chemical properties. Take CO_2-EOR for example; the CO_2-EOR process recovers oil that remains in the reservoir after primary and secondary recovery with the help of CO_2. Due to its special properties, CO_2 enhances oil recovery by lowering interfacial tension, swelling the oil, reducing oil viscosity, and mobilizing the lighter components of the oil [38]. Likewise, the other geological utilization approaches make use of CO_2 for enhanced gas/shale gas recovery, enhanced water recovery, enhanced coal bed methane recovery, and leaching of uranium, all of which are taking advantage of the properties of CO_2 and thus have the potential to utilize a large amount of CO_2 in the future (Figure 3.4). Geological utilization of CO_2 is highly related to CO_2 storage since approximately 60% injected CO_2 can be retained in the reservoir if reinjection is not considered [39]. It is therefore a significant component of CCUS technology. China utilizes 1.82 Mt CO_2 by geological CO_2 utilization per year [32]. PetroChina Jilin Oilfield EOR project is the largest EOR project in Asia, and more than 2 Mt of CO_2 have already been injected into this project [40]. In July 2021, Sinopec officially launched the first Mt-scale CCUS project in China (Qilu Petrochemical – Shengli Oilfield CCUS project). In August 2022, this Mt-scale CCUS project combining CO_2 capture and EOR was officially put into use, marking the CCUS industry in China entering a more mature stage of commercial application.

As shown in Figure 3.5, about 40 CCUS demonstration and pilot projects have been put into operation or under construction in China with a total capture capacity of 3 Mt per year [40, 41]. The predicted potential of CO_2 utilization technologies in China from 2025 to 2060 is given in Table 3.2. Compared with chemical or biological utilization, geological utilization can utilize larger amounts of CO_2 if the technology is mature enough and the cost is lower in the future.

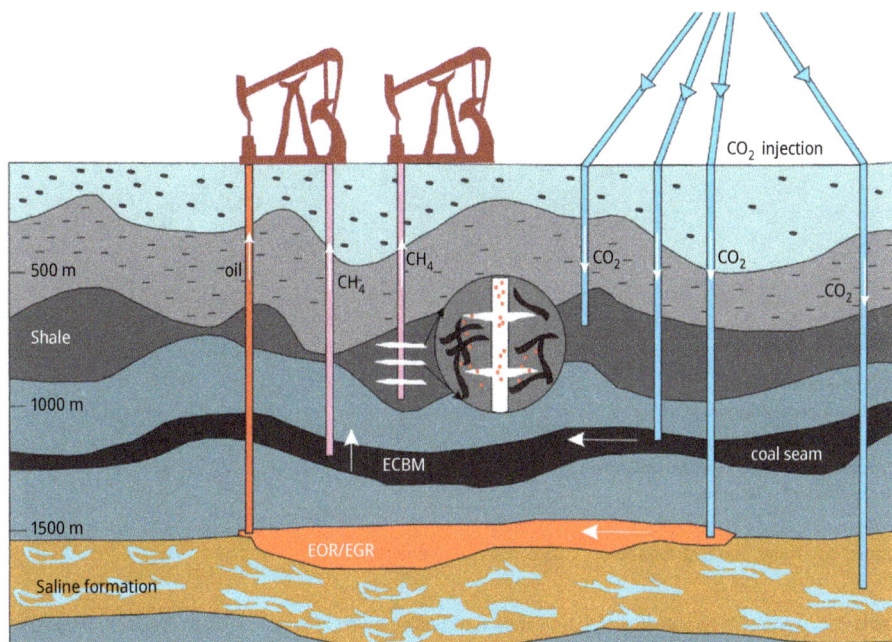

Figure 3.5: Schematic diagram of the geological CO_2 utilization technology in different geological reservoirs [42].

The comparison of the development of CO_2 utilization technologies between China and the other countries is shown in Figure 3.7. For chemical or biological utilization of CO_2 in China, most technologies have already achieved successful industrial demonstrations, and the technologies for CO_2 reforming to syngas, synthesis of organic carbonate, mineralization with steel slag or K-feldspar, and gas fertilizer are relatively matured in China. However, the research on synthesis of liquid fuels from CO_2 is relatively limited.

From Figure 3.6, even though China has made solid progress in CO_2 utilization, it is apparent that most technologies are still in the industrial demonstration stage. Both technology development and economic feasibility analysis are needed to brace for large-scale CO_2 utilization.

As shown in Figure 3.6, chemical and biological utilization of CO_2 can be mainly classified into two types based on the products: synthesis of chemicals/fuels, and syn-

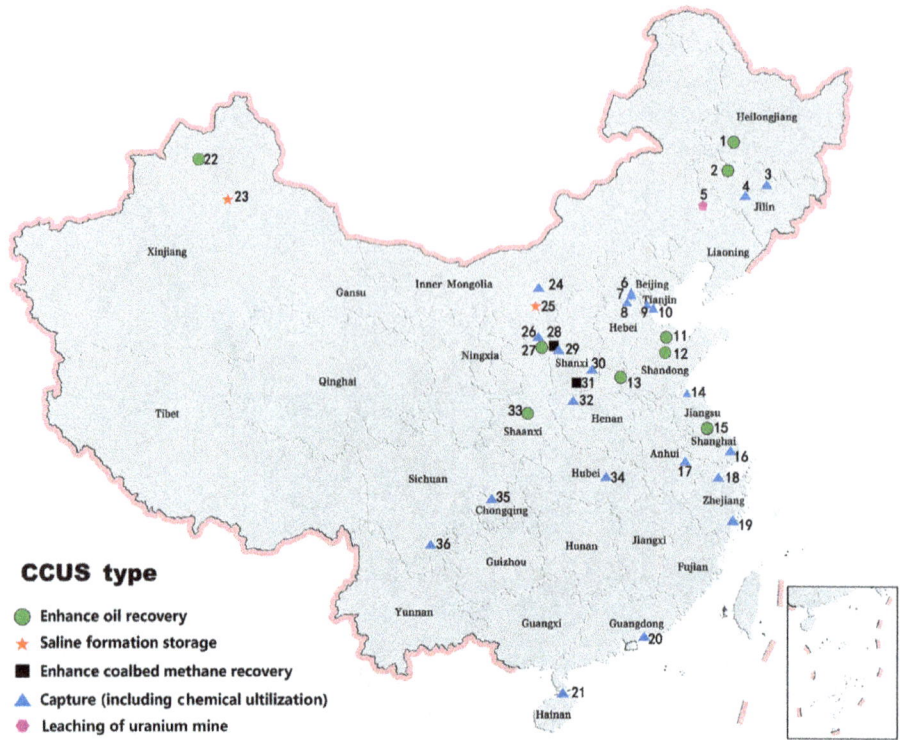

Figure 3.6: Types and distribution of CCUS projects in China [40, 43].

Table 3.2: The predicted potential of CO_2 utilization technologies in China from 2025 to 2060 (100 Mt/year) [32, 44, 45].

Year	2025	2030	2035	2040	2050	2060
Chemical or biological utilization	0.4–0.9	0.9–1.4	1.4–2.6	2.9–3.7	4.2–5.6	6.2–8.7
Geological utilization	0.1–0.3	0.5–1.4	1.3–4.0	3.3–8.0	5.4–14.3	6.0–20.5
Total	0.5–1.2	1.4–2.8	2.7–6.6	6.2–11.7	9.6–19.9	12.2–29.2

Note: The data of chemical or biological utilization cannot directly add up to the data of geological utilization due to different calculation methods.

thesis of materials. Common chemicals or fuel products include syngas, hydrocarbons, olefins, alcohols, acids, and aldehydes. The conventional way is thermal catalytic conversion of CO_2 with a coreactant such as H_2 or CH_4 [47]. However, this conventional technology requires a high temperature and thus has lower energy efficiency, which is a key issue affecting scale-up applications [48–50]. Hence, it is imperative to reduce the energy needed for CO_2 conversion to chemicals or fuels.

In addition to the conventional thermochemical conversion of CO_2, there are emerging pathways such as photochemical conversion [51–53], solar thermochemical

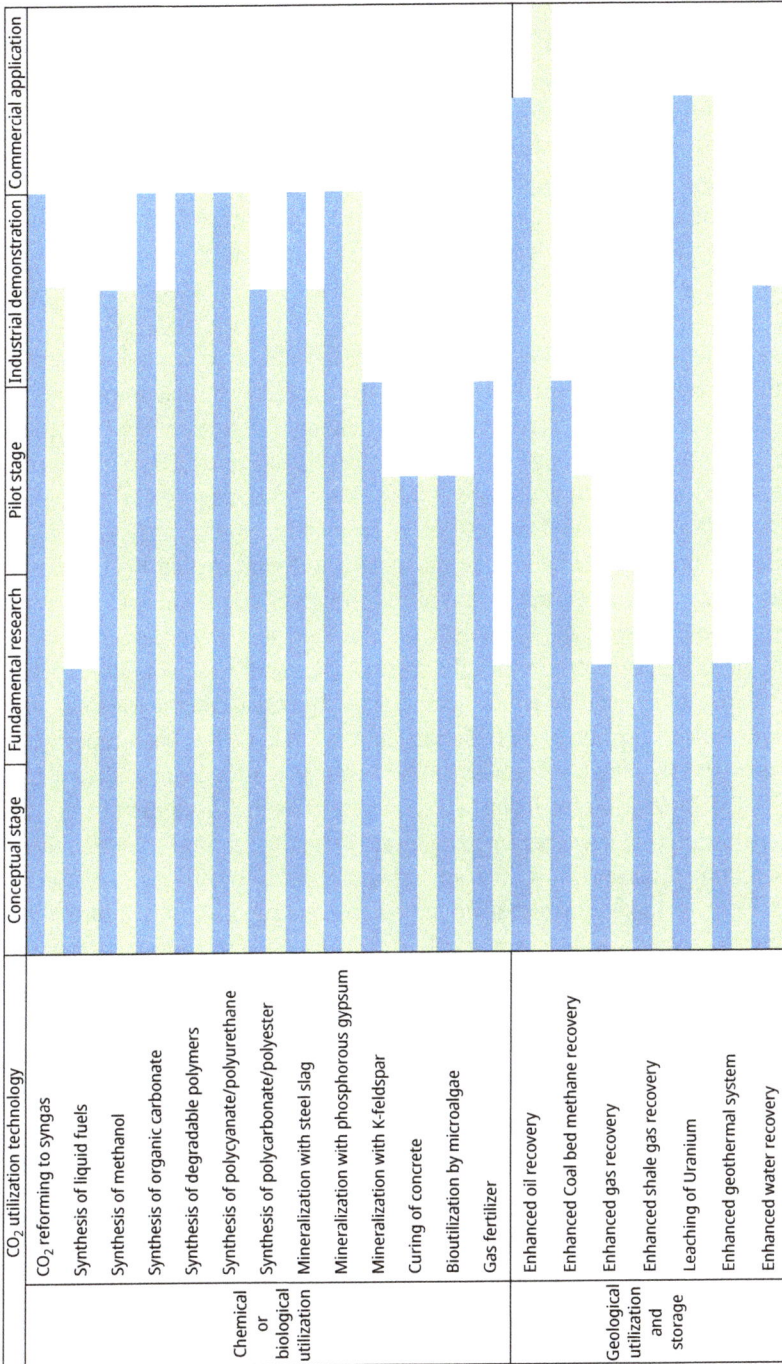

Figure 3.7: The comparison of the development of CO_2 utilization technologies between China and the other countries in 2022 (modified from [32, 46]).

conversion [54], electrochemical conversion [55–58], plasma-chemical conversion [59–62], and biochemical conversion [63, 64] that are noteworthy and therefore briefly discussed below. Solar-energy-driven conversion of CO_2 has attracted considerable interest worldwide, including solar thermochemical conversion and photocatalytic conversion of CO_2 [65]. For solar thermochemical conversion, concentrated solar radiation is used – in the form of high-temperature heat – as an energy source to drive the highly endothermic CO_2 dissociation reaction, whereas photochemical process uses the energy of a photon to excite electrons on photocatalysts and then reduces CO_2 to different products such as carbon monoxide, formaldehyde, formic acid, and alcohols [61].

As renewable energy-derived electricity becomes gradually more abundant and affordable, CO_2 conversion to value-added chemicals through electrochemical method and plasma-chemical method attracts increasing attention. Electrocatalytic CO_2 reduction is a promising strategy to convert CO_2 to chemicals [66]. Typical CO_2 electrolyzers reduce CO_2 to hydrocarbons and oxygenate at the cathode, whereas water is oxidized to oxygen at the anode [67]. Various chemicals can be produced including single-carbon (C_1) products, such as formic acid and carbon monoxide, and multicarbon (C_{2+}) products such as ethanol and ethylene. Plasma-chemical conversion of CO_2 to value-added products appears as another promising technology for more efficient large-scale CO_2 utilization since nonthermal plasma can initiate the thermodynamically unfavorable chemical reactions at near-ambient conditions (low temperature and atmospheric pressure) with a fast and high conversion rate [47, 61]. So far, plasma catalysis has been reported to successfully convert CO_2 to high-value fuels or chemicals such as methanol, ethanol, and acetic acid [47, 68]. Biological conversion is also an attractive approach to utilizing CO_2. Microalgae can convert solar energy into chemical energy by "natural" photosynthesis for the production of biofuels [61]. The advantages and disadvantages of these approaches are given in Table 3.3. To gear up large-scale applications in the future, these emerging technologies require deeper research to promote the conversion rate, enhance the catalyst lifespan and stability, tune the product selectivity, and increase the energy efficiency.

Various materials such as polymers, carbonates, and construction materials can be synthesized by chemical utilization of CO_2. CO_2 can be utilized as an alternative carbon source for the synthesis of polymers and carbonates [69]. The CO_2 curing of

---

Figure 3.7 (continued)

Note: 1. Blue color denotes the development level of China (2022) and green color denotes the average development level of the other countries (2022).

2. Conceptual stage means putting forward concept and idea; fundamental research means the completion of functional verification in a lab scale; pilot stage means the completion of the test in a medium scale; industrial demonstration means one to four industrial-scale operations are in progress or have completed tests; commercial application means five or more industrial-scale operations in progress or have completed tests.

concrete involves reactions between calcium silicate in cement and CO_2 in the presence of water to form both calcium carbonate and calcium silicate hydrate gel [70]. Since many industrial solid wastes such as steel slag, iron slag, cement-kiln dust, waste concrete, and coal flash are generally alkaline, inorganic, and rich in Ca or Mg, they are promising Ca or Mg resource for CO_2 carbonation to produce construction materials [65]. This technology utilizes solid industrial wastes, captures CO_2 together, and turns them into marketable products, and simultaneously undergoes CO_2 sequesteration within the infrastructure for a long time [71]. A technological challenge for the industrial-scale deployment of CO_2 carbonation is to accelerate the slow process of carbonation [65]. Besides, even though CO_2 carbonation can improve the durability of plain concrete-based structures, it will reduce the alkalinity of steel-reinforced materials and weaken the protection on steel bars [72]. Therefore, the development of corrosion-resistant materials, for instance, geopolymers, is needed for large-scale CO_2 utilizations via CO_2 carbonation in the future [73].

There are also technological issues to be tackled in geological utilization of CO_2. For example, compared with the CO_2-EOR status in the USA, the application of this technology in China may be more difficult as the geologic structure of most reservoirs is characterized by many faults and low permeability [74]. Hence, improving the sweep efficiency is the key to achieving extensive application of CO_2-EOR in China [42]. Likewise, the other geological approaches for CO_2 utilization need to address the technology bottlenecks caused by the domestic geological features. In addition, uncertainty related to the long-term underground behavior of CO_2 is a concern and requires monitoring [69].

Table 3.3: Advantages and disadvantages of various chemical or biological approaches for CO_2 utilization to produce fuels or chemicals.

Approaches	Advantages	Disadvantages	References
Thermal	– High conversion – High yield	– Low energy efficiency – Low material stability	[55, 75]
Solar thermochemical	– Direct utilization of sunlight – Easy to scale up	– Low material stability – Limited products	[54, 76]
Photochemical	– Do not require additional energy – Environment friendly – Economically feasible	– Low conversion – Produce a less yield	[77–79]
Electrochemical	– No additional heat is required – Easy to scale up – Renewable energy can be utilized – High selectivity	– Low catalyst stability – Kinetic barriers – Need post-reaction separation processes	[55, 65, 80]

Table 3.3 (continued)

Approaches	Advantages	Disadvantages	References
Plasma-chemical	– Easy to scale up – Fast and high conversion – Renewable energy can be utilized – High flexibility	– Need post-reaction separation processes – Low selectivity to liquid products	[47, 61, 68, 81]
Biological	– Toxic tolerance – Higher selectivity – Require ambient conditions	– Time-consuming – High cost for cultivation – Need biorefinery	[63, 69, 82]

Economic feasibility is another critical factor for successful large-scale CO_2 utilization. CO_2 utilization should be integrated with CO_2 capture technology (CCUS) in large-scale applications, where it is also likely to be combined with CO_2 transportation and CO_2 storage (CCUS). Relevant technoeconomic analysis (TEA) has suggested the economic feasibility of different CCUS projects in China [83–85]. For instance, TEA research on a 1 million-scale CCUS-EOR project in China indicates that the internal rate of return can reach 12.9% under different scenarios over the 20-year project period; and according to the sensitivity analysis, oil price and the financial support from the government are the most influential economic factors [83]. However, the cost of CCUS is still too high. In addition to the fixed cost from construction and equipment installation, the cost for CCUS projects comes from CO_2 capture, CO_2 transportation, as well as CO_2 storage, and CO_2 capture turns out to be the major economic contributor. The energy consumption and economic cost of CO_2 capture depend on the source of CO_2 and the CO_2 concentration [41, 86]. Technically, the higher the CO_2 concentration from the source, the lower the energy consumption and economic cost. The development of novel CO_2 capture technologies could minimize energy penalty and reduce the cost of energy-intensive separation of CO_2 from flue gases and industrial streams [87]. In the long run, cheaper direct air capture (based on clean energy) could also support the CO_2 utilization [88]. Influenced by the maturity of the technology in China, the CO_2 capture cost in cement industry and electric power industry is quite high, while the CO_2 capture cost in coal chemical industry and petrochemical industry is relatively lower [41]. In the future, China needs to further reduce the cost of CCUS and promote the industrial demonstrations of different technologies.

3.3 Hydrogen energy

3.3.1 The policies on hydrogen production in China

Responding to CE and carbon neutrality goals, the action on building a low-carbon society and a nonfossil energy supply system has become increasingly important. Electrification is of particular importance to the energy transition to carbon neutrality [37]. Hence, it is essential to make more active use of renewable sources of energy (e.g., solar energy, wind, hydro, geothermal, and biomass) and design better conversion systems without negative environmental impacts [20]. However, renewable energy development may have some negative impacts on the human and ecological environments (effects of solar energy facilities on land use, biomass growth on land area, wind power on birds, hydropower on aqua life, etc.). In addition, the use of renewable energy sources encounters issues of regional distribution, seasonable variability, and even daily change as well as energy density issues. Hydrogen, a zero-carbon fuel with a high gravimetric energy density, is therefore enjoying unprecedented political and business momentum. It is now regarded as one key element of a potential energy solution for the twenty-first century, capable of assisting in major issues of CO_2 emissions, sustainability, and energy security [89, 90]. The potential of hydrogen energy in CO_2 emissions reduction is shown in Figure 3.8 [37]. Heavy industry and long-distance transportation contribute to 80% of emission reductions from the use of low-carbon hydrogen and hydrogen-rich fuels. In addition, hydrogen is crucial as a coreactant of CO_2 utilization to synthesize high-value chemicals or fuels [47]. Hence, hydrogen energy is promising as a necessary component of China's future clean energy landscape.

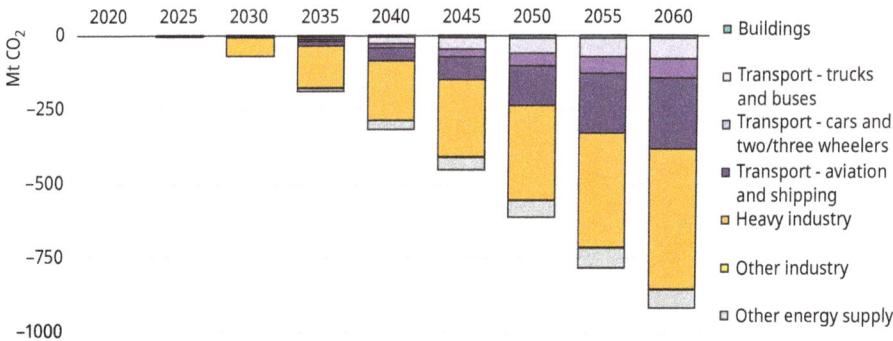

Figure 3.8: The prediction for avoided CO_2 emissions by using hydrogen energy in China [37].

Unlike petroleum or natural gas, H_2 molecule does not exist in nature as a resource, and must be produced by using sources of hydrogen atom with energy input [20]. Hydrogen can be produced from many different renewable and nonrenewable feedstocks and technological pathways, with quite different greenhouse gas emissions [91]. However, from the perspective of CE and carbon neutrality, it is essential to demonstrate that hydrogen production has sufficiently low carbon emissions. According to the type of energy source or additional technology utilized to produce that hydrogen, hydrogen may be labeled by three main colors: gray, blue, and green (Table 3.4). Gray hydrogen is regarded as a type of hydrogen whose production is accompanied by pollution, while blue hydrogen defines the CO_2 capture inclusive of gray hydrogen. Green hydrogen refers to a 100% renewable energy source for hydrogen production [92]. Hence, green hydrogen, a clean secondary energy as well as an efficient energy carrier, is a potential source of energy in line with carbon neutrality goals.

Table 3.4: The difference among gray, blue, and green hydrogen (from IRENA) [93].

Color	Gray	Blue	Green
Process	SMR (steam methane reforming) or gasification	SMR or gasification with carbon capture	Electrolysis of water
Source	Methane or coal	Methane or coal	Renewable electricity

China has attached great significance to hydrogen energy in recent years, aiming at achieving green and low-carbon energy transition. Hydrogen energy is first included in government report in 2019 [94]. Since then, more and more policies have been issued to support the development of hydrogen production. Nonetheless, the major source for hydrogen production in China is still fossil energy such as coal (64%) and natural gas (14%), only about 1% come from electrolysis of water [95]. As shown in Figure 3.9, the total hydrogen production in 2020 is 33.4 Mt, among which green hydrogen production is only around 0.5 Mt [96]. The major portion of hydrogen (84.1%) is currently used in the industry for chemical processes such as methanol synthesis and ammonia synthesis. With the promotion by the government and the development of relevant technologies, the ratio of green hydrogen in hydrogen production is predicted to increase to 15% by 2030 [97]. In 2060, the total amount of hydrogen production is predicted to be 130 Mt, among which 81% is from green hydrogen and 15% from blue hydrogen [96]. The main consumption of hydrogen will still be the industry (~60%), followed by transportation system (~30%). Hence, green hydrogen will gradually become an important component of China's new energy supply and consumption system. The recent relevant policies in China explicitly favor green hydrogen over gray hydrogen and promote the development of green hydrogen production (Table 3.5).

(a) Data in 2020

(b) Predicted data in 2060

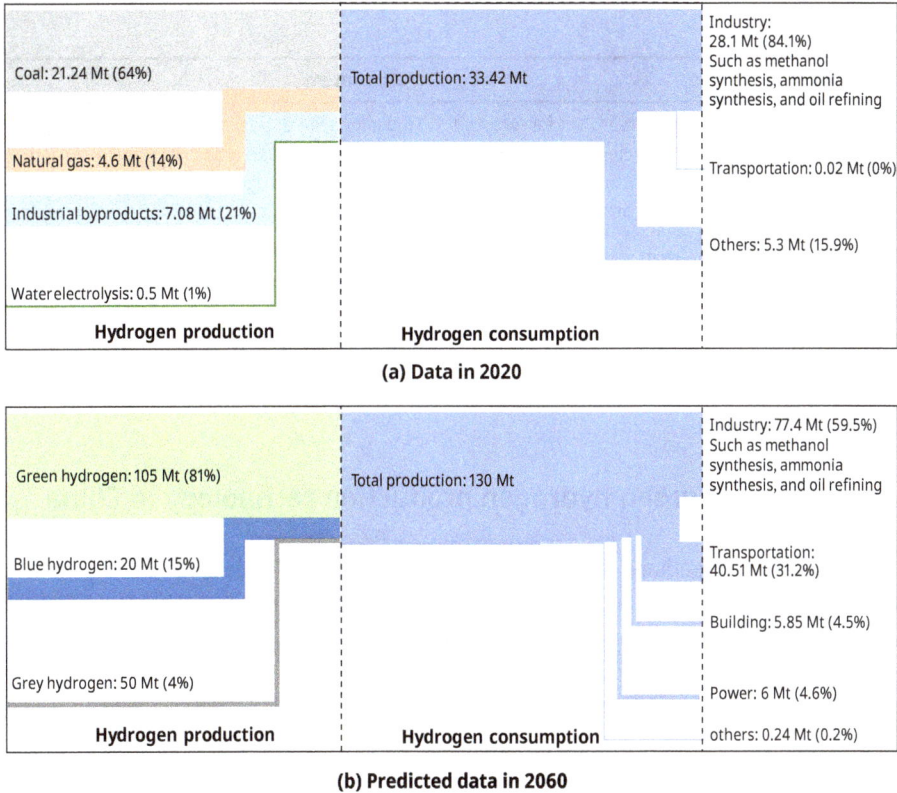

Figure 3.9: The current (a) and predicted (b) data for hydrogen production and consumption in China (modified from [96]).

Table 3.5: Recent policies on hydrogen energy in China [98, 99].

Time	Policy	Main content
2019	Government Work Report 2019	To promote the construction of charging and hydrogen refueling facilities
2020	Opinions on Accelerating the Establishment of Green Production and Consumption Laws and Policies	The promotion of clean energy development requires the study and formulation of standards and supporting policies for new technologies, such as hydrogen energy
2021	Guiding Opinions on Accelerating the Establishment and Improvement of the Green and Low-Carbon Circular Development Economic Systems	To promote the green and low-carbon transformation of the energy system and develop hydrogen energy

Table 3.5 (continued)

Time	Policy	Main content
2021	Outline of the 14th Five-Year Plan for National Economic and Social Development and the Long-Range Objectives Through the Year 2035 for the People's Republic of China	To deploy several future industries in cutting-edge technology and industrial transformation fields such as hydrogen energy and energy storage
2022	Medium- and long-term plan for the development of the hydrogen energy industry (2021–2035)	To put in place a relatively complete hydrogen energy industry development system by 2025; the proportion of hydrogen produced from renewable energy in terminal energy consumption will increase significantly by 2035

3.3.2 Status of green hydrogen production technology in China

There are several methods for green hydrogen production (Table 3.6), including water electrolysis, H_2S cracking, and biomass conversion. Among them, water electrolysis is the major method to produce green hydrogen in China. The alkaline water electrolysis

Table 3.6: Production methods and energy conversion paths to generate green hydrogen [101].

Hydrogen production method	Green energy source
Electrolysis or plasma arc decomposition	Electricity generated from solar/geothermal/biomass/wind/ocean heat, and so on
Thermocatalysis for H_2S cracking	Concentrated solar heat/biomass combustion
Thermochemical biomass conversion	Heat generated from solar/geothermal/biomass
Thermochemical water splitting	
Thermochemical gasification	
Thermochemical fuel reforming	
Thermochemical H_2S splitting	
PV electrolysis	Solar energy
Photocatalysis	Solar energy
Photoelectrolysis	Solar energy
Biophotolysis	Biomass and photonic energy
Dark fermentation	Biomass
Artificial photosynthesis	Solar energy

process and the proton exchange membrane (PEM) process are two relatively mature technologies for water electrolysis in China (Table 3.7). Solid oxide electrolyzer cell and anion exchange membrane are emerging approaches with high energy efficiency and product purity. However, they are still in the stage of fundamental research and much remains to be developed before being commercialized [100]. In the coming years, alkaline setup will continue to be the major technology for green H_2 production in China, and PEM is estimated to reach 10% market share in water electrolysis in 2030 [97].

Table 3.7: Comparison between the alkaline process and the proton exchange membrane process [97, 102].

	Alkaline	PEM
Electrolyte	KOH/NaOH 25–30 wt%	Polymer (NAFION)
Ion transport	OH^-	H^+
H_2 product purity	$\geq 99.8\%$	$\geq 99.99\%$
Current density (A/cm^2)	<0.8	1–4
Energy efficiency (%)	60–75	70–90
Energy consumption (kWh/Nm^3)	4.5–5.5	4.0–5.0
Cell cost (USD/kW)	400–600	2000
R&D level	Fully commercialized	Preliminarily commercialized

Nonetheless, the current cost for hydrogen production from water electrolysis is still much higher than that from coal or natural gas. According to IEA, only if the cost for hydrogen production is reduced to lower than US$3/kg can it be cost-competitive with the conventional production methods (from coal or natural gas) [103], while US DOE has set a target of hydrogen production cost at US$2/kg-$H_2$. From an economic feasibility analysis [104, 105], the cost for hydrogen production by alkaline process (alkaline water electrolysis) is about RMB 30/kg (~US$4.3/kg), mainly from electricity cost (74%) and the depreciation of equipment (18%). On the one hand, for every RMB 0.1/kWh decrease in electricity cost, the hydrogen cost decreases by RMB 0.5/Nm^3 on average. On the other hand, increasing the utilization of the equipment can also reduce the cost to some extent. The cost for hydrogen production by the alkaline process is predicted to drop by 5–10% with the development of related technologies and the scale expansion in the next 10 years.

Compared with the alkaline process, the cost for hydrogen production by PEM is higher at about RMB 40/kg (~US$4.7/kg), mainly due to the high cost of the PEM cell [104]. Therefore, the depreciation of equipment accounts for 44% of the total cost, while the electricity cost accounts for 50%. Because of the technology limitations on PEM, the cost of hydrogen production by PEM will continue to be higher than the alkaline process in the medium term (by 2030). However, the space requirement of PEM equipment is almost half of that of alkaline equipment under the same H_2 production scale, so the advantage of PEM will be more distinct in the area where the land is expensive. Moreover, PEM is more compatible with the wind or photovoltaic (PV) power generation system due to its high workload capacity and fast response speed [102, 106, 107]. Consequently, PEM is a promising technology for green hydrogen production in the long run.

Since green hydrogen production has a great demand for renewable-based electricity, there are several challenges in constructing green hydrogen supply system in China. One issue is the spatial mismatch between green hydrogen supply and demand [96]. Solar PV and wind power will be the main energy sources for hydrogen production in China. Large wind power bases are concentrated in the northwestern and northern regions, and offshore wind power bases are mainly distributed in the southeastern coastal areas. However, a major part of hydrogen demand is from the central and eastern regions. Another issue is the mismatch between the green hydrogen supply system and existing policies and standards [96]. For instance, the current regulations and standards for hydrogen energy mainly focus on fuel cells and transportation applications. The standard system for green hydrogen supply lags behind the development of the industry, lacking the support of engineering data and demonstration cases.

Hence, the further development of green hydrogen production in China needs not only the technology support but also the policy support. From the technology perspective, hydrogen storage and transportation technologies are of vital importance in terms of the spatial mismatch issue as mentioned above [108]. The other important directions include the development of better catalysts, the optimization of reactor design, the deepening of research on hydrogen safety, and the improvement of manufacturing process. Moreover, the development of power generation technology from renewable energy will be conducive for large-scale application for hydrogen production. From the policy perspective, a systematic policy framework with detailed roadmaps is needed.

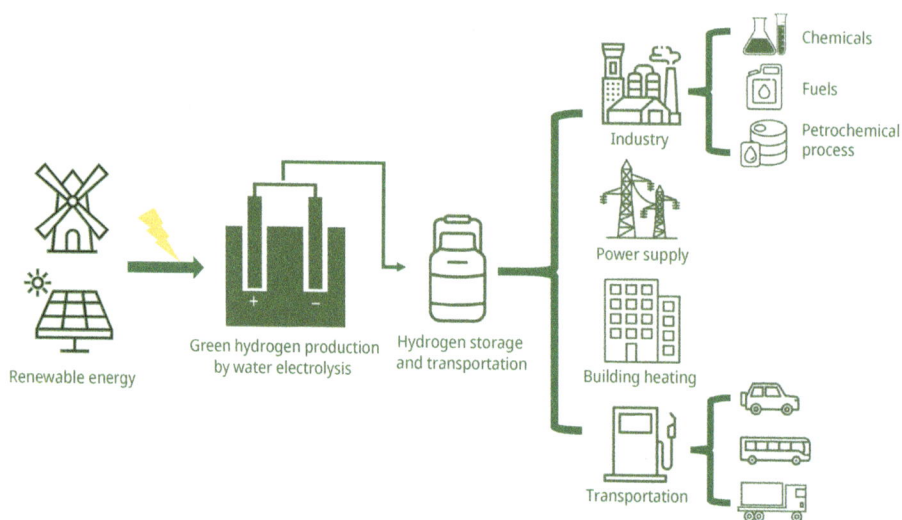

Figure 3.10: The industrial chain of green hydrogen energy in the future (modified from [100]).

The government is establishing a "1 + N" policy framework for the development of H_2 energy in China, aiming at promoting the industrial demonstrations and providing more financial support for the industries [109]. In the future, China needs to further establish and improve the relevant top-level design and standard system for the production, storage, transportation, and management of hydrogen as the energy product, and improve the safety for the overall hydrogen energy system [105]. The industrial chain of green hydrogen energy in the future is shown in Figure 3.10. Green hydrogen has the potential to be the core of the future industrial system, and leads to a low-carbon society in China.

3.4 Conclusion

With the development of CE in China in recent decades, the scope of the CE is no longer limited to waste management but also covers a broader range in line with tackling climate change. The all-round CE framework in China forms an important component of its broader climate strategy to tackle climate change and achieve carbon neutrality by 2060. China formulates and implements a "1 + N" policy framework to support the carbon neutrality goal. CO_2 utilization plays a key role in achieving carbon neutrality goals in that it captures CO_2 as a resource and then reuses it to close the carbon cycle. The green hydrogen produced by 100% renewable energy is a prospective and efficient energy carrier in line with CE and carbon emission reduction. This chapter briefly summarized the policies and the state-of-the-art technologies for both CO_2 utilization and green hydrogen production. To brace for large-scale applications, research and development, technological efforts, policy support, and government supervision are all needed for reaching a CE and carbon neutrality:

(1) For various routes of chemical and biological utilization of CO_2, reducing the energy consumption and finding a tailored catalyst (catalytic material) and reaction process for target reactions are important directions for research and technology development.

(2) For geological utilization of CO_2, more efforts are needed to address the technology bottlenecks caused by the domestic geological features of China.

(3) The CCUS projects in China are potentially feasible under the target of carbon neutrality but still requires cost reduction especially in CO_2 capture and CO_2 conversion and utilization.

(4) The green hydrogen energy will be more competitive with further decrease in the cost of renewable-based electricity and is a promising component in the future industrial system.

(5) PEM is a promising technology for green hydrogen production via water electrolysis in the long run. Direct use of seawater would be more desirable if it becomes feasible.

(6) A systematic and detailed policy framework regarding hydrogen as an energy product is needed for large-scale hydrogen production in China related to the target of carbon neutrality.

References

[1]   MacArthur, E., Towards the circular economy. Journal of Industrial Ecology, 2013, 2(1), 23–44.
[2]   Geissdoerfer, M., et al. The circular economy – a new sustainability paradigm?. Journal of Cleaner Production, 2017, 143, 757–768.
[3]   Valko, D., Circular economy: A theoretical model and implementation effects. National Interests: Priorities and Security, 2018, 14(8), 1415–1419.
[4]   Ilinova, A. and Kuznetsova, E., CC (U) S initiatives: Prospects and economic efficiency in a circular economy. Energy Reports, 2022, 8, 1295–1301.
[5]   Ghosh, S.K., Circular Economy: Global Perspective. Springer, 2020.
[6]   Liu, L. and Ramakrishna, S., An Introduction to Circular Economy. Springer, 2021.
[7]   Zhu, J., et al. Efforts for a circular economy in China: A comprehensive review of policies. Journal of Industrial Ecology, 2019, 23(1), 110–118.
[8]   Su, B., et al. A review of the circular economy in China: Moving from rhetoric to implementation. Journal of Cleaner Production, 2013, 42, 215–227.
[9]   Shi, L., et al. Circular economy: A new development strategy for sustainable development in China. 3rd World Congress of Environmental and Resource Economists, 2006, pp. 3–7.
[10] McDowall, W., et al. Circular economy policies in China and Europe. Journal of Industrial Ecology, 2017, 21(3), 651–661.
[11] Bleischwitz, R., et al. The circular economy in China: Achievements, challenges and potential implications for decarbonisation. Resources, Conservation and Recycling, 2022, 183, 106350.
[12] Fan, Y. and Fang, C., Circular economy development in China-current situation, evaluation and policy implications. Environmental Impact Assessment Review, 2020, 84, 106441.
[13] Qi, J., et al. The role of government and China's policy system for circular economy, Singapore. Development of Circular Economy in China. Springer, 2016, 21–53.
[14] Feng, K. and Lam, C.-Y., An overview of circular economy in China: How the current challenges shape the plans for the future. The Chinese Economy, 2021, 54(5), 355–371.
[15] PRC, N.d.a.r.c., The development plan for the circular economy in the 14th five year plan period (2021–2025). 2021. Available from: https://www.ndrc.gov.cn/xxgk/zcfb/ghwb/202107/P020210707324072693362.pdf.
[16] Dong, H., et al. Uncovering energy saving and carbon reduction potential from recycling wastes: A case of Shanghai in China. Journal of Cleaner Production, 2018, 205, 27–35.
[17] Wei, W., et al. Toward carbon neutrality: Circular economy approach and policy implications. Bulletin of Chinese Academy of Sciences, 2021, 36(9), 1030–1038.
[18] Geng, Y. and Doberstein, B., Developing the circular economy in China: Challenges and opportunities for achieving 'leapfrog development'. The International Journal of Sustainable Development & World Ecology, 2008, 15(3), 231–239.
[19] NOAA. Trends in atmospheric carbon dioxide. 2023. Available from: https://gml.noaa.gov/ccgg/trends/.

[20]   Song, C., Global challenges and strategies for control, conversion and utilization of CO_2 for sustainable development involving energy, catalysis, adsorption and chemical processing. Catalysis Today, 2006, 115(1–4), 2–32.

[21]   IEA, *Global Energy Review 2021*. 2021.

[22]   Kjellstrom, T., et al. Heat, human performance, and occupational health: A key issue for the assessment of global climate change impacts. Annual Review of Public Health, 2016, 37(1), 97–112.

[23]   Miraglia, M., et al. Climate change and food safety: An emerging issue with special focus on Europe. Food and Chemical Toxicology, 2009, 47(5), 1009–1021.

[24]   Shan, Y., et al. China CO_2 emission accounts 2016–2017. Scientific Data, 2020, 7(1), 1–9.

[25]   Wen, L., et al. Carbon emissions from multiple industries in China under the background of carbon peak and carbon neutrality and mitigation strategies. Statistics and Application, 2022, 11, 1143.

[26]   Liu, Z., et al. Challenges and opportunities for carbon neutrality in China. Nature Reviews Earth & Environment, 2022, 3(2), 141–155.

[27]   PRC, N.d.a.r.c. Working Guidance for Carbon Dioxide Peaking and Carbon Neutrality in Full and Faithful Implementation of the New Development Philosophy. 2021. Available from: https://en.ndrc. gov.cn/policies/202110/t20211024_1300725.html.

[28]   PRC, N.d.a.r.c., ACTION PLAN FOR CARBON DIOXIDE PEAKING BEFORE 2030. 2021. Available from: https://en.ndrc.gov.cn/policies/202110/t20211027_1301020.html.

[29]   PRC, N.d.a.r.c. The Sci-tech action plan to support carbon peaking and carbon neutrality (2022–2030). 2022; Available from: https://www.gov.cn/zhengce/zhengceku/2022-08/18/5705865/files/ 94318119b8464e2583a3d4284df9c855.pdf.

[30]   Li, X., et al. Early opportunities of carbon capture and storage in China. Energy Procedia, 2011, 4, 6029–6036.

[31]   Jiutian, Z., et al. Several key issues for CCUS development in China targeting carbon neutrality. Carbon Neutrality, 2022, 1(1), 1–20.

[32]   Cai, B., et al. China Status of CO_2 Capture, Utilization and Storage (CCUS) 2021 — China's CCUS pathways. Chinese Academy of Environmental Planning, 2021, 8.

[33]   Alsarhan, L.M., et al. Circular carbon economy (CCE): A way to invest CO_2 and protect the environment, a review. Sustainability, 2021, 13(21), 11625.

[34]   Zhang, X., et al. Catalytic conversion of carbon dioxide to methanol: Current status and future perspective. Frontiers in Energy Research, 2021, 8, 621119.

[35]   Song, C., *CO_2 conversion and utilization: An overview*, in *CO_2 conversion and utilization*. American Chemical Society, 2002, 2–30.

[36]   Ormerod, W., Riemer, P. and Smith, A., Carbon Dioxide Utilisation. IEA Greenhouse Gas R&D Programme. IEA, London, UK. 1995.

[37]   IEA. An Energy Sector Roadmap to Carbon Neutrality in China. IEA Paris, 2021.

[38]   Verma, M.K., *Fundamentals of Carbon Dioxide-Enhanced Oil Recovery (CO_2-EOR): A Supporting Document of the Assessment Methodology for Hydrocarbon Recovery Using CO_2-EOR Associated with Carbon Sequestration*. 2015: US Department of the Interior, US Geological Survey. Washington, DC.

[39]   Gozalpour, F., Ren, S.R. and Tohidi, B., CO_2 EOR and storage in oil reservoir. Oil & Gas Science and Technology, 2005, 60(3), 537–546.

[40]   Liu, G., et al. China's pathways of CO_2 capture, utilization and storage under carbon neutrality vision 2060. Carbon Management, 2022, 13(1), 435–449.

[41]   Zhang, X., et al. Development of carbon capture, utilization and storage technology in China. Strategic study of CAE, 2021, 23(6), 70–80.

[42]   Liu, H., et al. Worldwide status of CCUS technologies and their development and challenges in China. Geofluids, 2017, 2017.

[43] Cai, B., et al. *China Status of CO₂ Capture, Utilization and Storage (CCUS) 2019*. Center for Climate Change and Environmental Policy, Chinese Academy of Environmental Planning. Beijing, China. 2020, 1–200316151449.
[44] Wei, N., et al. Regional resource distribution of onshore carbon geological utilization in China. Journal of CO₂ Utilization, 2015, 11, 20–30.
[45] Qin, J., et al. CCUS: Global progress and China's policy suggestions. Petroleum Geology and Recovery Efficiency, 2020, 27(1), 20–28.
[46] Ministry of Science and Technology of PRC., et al. Roadmap for carbon capture, utilization and storage technology development in China (2019). Beijing: Science Press, 2019.
[47] Ray, D., et al. Recent progress in plasma-catalytic conversion of CO₂ to chemicals and fuels. Catalysis Today, 2022, *423*: 113973.
[48] Satthawong, R., et al. Bimetallic Fe–Co catalysts for CO₂ hydrogenation to higher hydrocarbons. Journal of CO₂ Utilization, 2013, 3, 102–106.
[49] Jiang, X., et al. Bimetallic Pd–Cu catalysts for selective CO₂ hydrogenation to methanol. Applied Catalysis B: Environmental, 2015, 170, 173–185.
[50] Wang, J., et al. CO₂ hydrogenation to methanol over In₂O₃-based catalysts: From mechanism to catalyst development. ACS Catalysis, 2021, 11(3), 1406–1423.
[51] Ran, J., Jaroniec, M. and Qiao, S.Z., Cocatalysts in semiconductor-based photocatalytic CO₂ reduction: Achievements, challenges, and opportunities. Advanced Materials, 2018, 30(7), 1704649.
[52] Nahar, S., et al. Advances in photocatalytic CO₂ reduction with water: A review. Materials, 2017, 10(6), 629.
[53] Albero, J., Peng, Y. and García, H., Photocatalytic CO₂ reduction to C2+ products. ACS Catalysis, 2020, 10(10), 5734–5749.
[54] Pullar, R.C., et al. A review of solar thermochemical CO₂ splitting using ceria-based ceramics with designed morphologies and microstructures. Frontiers in Chemistry, 2019, 7:601.
[55] Saravanan, A., et al. A comprehensive review on different approaches for CO₂ utilization and conversion pathways. Chemical Engineering Science, 2021, 236, 116515.
[56] Whipple, D.T. and Kenis, P.J., Prospects of CO₂ utilization via direct heterogeneous electrochemical reduction. The Journal of Physical Chemistry Letters, 2010, 1(24), 3451–3458.
[57] Xie, Y., et al. High carbon utilization in CO₂ reduction to multi-carbon products in acidic media, *5*(6): 564–570. Nature Catalysis, 2022, 1–7.
[58] Zhang, Z., Xie, Y. and Wang, Y., What matters in the emerging application of CO₂ electrolysis. Current Opinion in Electrochemistry, 2022, 34: 101012.
[59] Wang, J., et al. One-step plasma-enabled catalytic carbon dioxide hydrogenation to higher hydrocarbons: Significance of catalyst-bed configuration. Green Chemistry, 2021, 23(4), 1642–1647.
[60] Yao, X., et al. Plasma-catalytic conversion of CO₂ and H₂O into H₂, CO, and traces of CH₄ over NiO/cordierite catalysts. Industrial & Engineering Chemistry Research, 2020, 59(43), 19133–19144.
[61] Snoeckx, R. and Bogaerts, A., Plasma technology – A novel solution for CO₂ conversion?. Chemical Society Reviews, 2017, 46(19), 585–5863.
[62] Wang, J., et al. Synergetic effect of nonthermal plasma and supported cobalt catalyst in plasma-enhanced CO₂ hydrogenation. Chemical Engineering Journal, 2022, 451: 138661.
[63] Yaashikaa, P., et al. A review on photochemical, biochemical and electrochemical transformation of CO₂ into value-added products. Journal of CO₂ Utilization, 2019, 33, 131–147.
[64] Gupta, R., et al. Biochemical conversion of CO₂ in fuels and chemicals: Status, innovation, and industrial aspects. Biomass Conversion and Biorefinery, 2022, 1–24, *14*(3): 3007–3030.
[65] Zhu, Q., Developments on CO₂-utilization technologies. Clean Energy, 2019, 3(2), 85–100.
[66] Usman, M., et al. Electrochemical reduction of CO₂: A review of cobalt based catalysts for carbon dioxide conversion to fuels. Nanomaterials, 2021, 11(8), 2029.

[67] Overa, S., et al. Electrochemical approaches for CO_2 conversion to chemicals: A journey toward practical applications. Accounts of Chemical Research, 2022, 55(5), 638–648.

[68] Ashford, B. and Tu, X., Non-thermal plasma technology for the conversion of CO_2. Current Opinion in Green and Sustainable Chemistry, 2017, 3, 45–49.

[69] Meylan, F.D., Moreau, V. and Erkman, S., CO_2 utilization in the perspective of industrial ecology, an overview. Journal of CO_2 Utilization, 2015, 12, 101–108.

[70] Monkman, S. Carbon dioxide utilization in fresh industrially produced ready mixed concrete. in *International Concrete Sustainability Conference*. 2014.

[71] Hepburn, C., et al. The technological and economic prospects for CO_2 utilization and removal. Nature, 2019, 575(7781), 87–97.

[72] Zhang, N., et al. Utilization of CO_2 into recycled construction materials: A systematic literature review. Journal of Material Cycles and Waste Management, 2022, 1–18.

[73] Goyal, A., et al. A review of corrosion and protection of steel in concrete. Arabian Journal for Science and Engineering, 2018, 43(10), 5035–5055.

[74] Yue, X., Zhao, R. and Zhao, F., Technological challenges for CO_2 EOR in China. Science Paper Online, 2007, 2(7), 487–491.

[75] Hu, B., Guild, C. and Suib, S.L., Thermal, electrochemical, and photochemical conversion of CO_2 to fuels and value-added products. Journal of CO_2 Utilization, 2013, 1, 18–27.

[76] Mustafa, A., et al. Current technology development for CO_2 utilization into solar fuels and chemicals: A review. Journal of Energy Chemistry, 2020, 49, 96–123.

[77] Gao, Y., et al. Recent advances in visible-light-driven conversion of CO_2 by photocatalysts into fuels or value-added chemicals. Carbon Resources Conversion, 2020, vol. 3, 46–59.

[78] Li, K., et al. A Critical Review of CO_2 Photoconversion: Catalysts and Reactors. Catalysis Today, 2014, vol. 224, 3–12.

[79] Alper, E. and Orhan, O.Y., CO_2 utilization: Developments in conversion processes. Petroleum, 2017, 3(1), 109–126.

[80] Agarwal, A.S., et al. *Technology Development for Large Scale Electrochemical Conversion of CO_2 to Useful Products*. In *Proceedings of the Clean Technology 2011 Conference & Expo, Boston*, conference article, (pp. 13–16), MA, USA. 2011.

[81] Bogaerts, A., et al. The 2020 plasma catalysis roadmap. Journal of Physics D: Applied Physics, 2020, 53(44), 443001.

[82] Appel, A.M., et al. Frontiers, opportunities, and challenges in biochemical and chemical catalysis of CO_2 fixation. Chemical Reviews, 2013, 113(8), 6621–6658.

[83] Wang, Y. Techno-economic assessment of a full-chain CCUS-EOR project based on carbon capture of iron/steel industry. Beijing: North China Electric Power University(Beijing), 2018.

[84] Gu, Y., et al. Techno-economic analysis of green methanol plant with optimal design of renewable hydrogen production: A case study in China. International Journal of Hydrogen Energy, 2022, 47(8), 5085–5100.

[85] Lin, H., et al. Techno-economic evaluation of coal-based polygeneration systems of synthetic fuel and power with CO_2 recovery. Energy Conversion and Management, 2011, 52(1), 274–283.

[86] Chen, C., Kotyk, J.F.K. and Sheehan, S.W., Progress toward commercial application of electrochemical carbon dioxide reduction. Chem, 2018, 4(11), 2571–2586.

[87] Wang, X. and Song, C., Carbon capture from flue gas and the atmosphere: A perspective. Frontiers in Energy Research, 2020, 8, 560849.

[88] Keith, D.W., et al. A process for capturing CO_2 from the atmosphere. Joule, 2018, 2(8), 1573–1594.

[89] Edwards, P.P., Kuznetsov, V. and David, W.I., *Hydrogen energy*. Philosophical Transactions of the Royal Society A: Mathematical, Physical and Engineering Sciences, 2007, 365(1853), 1043–1056.

[90] Momirlan, M. and Veziroglu, T., Current status of hydrogen energy. Renewable and Sustainable Energy Reviews, 2002, 6(1–2), 141–179.

[91] Abad, A.V. and Dodds, P.E., Green hydrogen characterisation initiatives: Definitions, standards, guarantees of origin, and challenges. Energy Policy, 2020, 138, 111300.

[92] Dawood, F., Anda, M. and Shafiullah, G., Hydrogen production for energy: An overview. International Journal of Hydrogen Energy, 2020, 45(7), 3847–3869.

[93] IRENA. *Policies for green hydrogen*. Available from: https://www.irena.org/Energy-Transition/Policy/Policies-for-green-hydrogen.

[94] PRC, T.G.O., *Report on the work of the government* 2019. Available from: http://www.gov.cn/premier/2019-03/16/content_5374314.htm.

[95] Liu, W., et al. Key technology of water electrolysis and levelized cost of hydrogen analysis under carbon neutral vision. Electrical Engineering magazine, 2022, 37(11), 2888–2896.

[96] Du, Z., et al. Construction of green-hydrogen supply system in China: reflections and suggestions. Strategic study of CAE, 2022, 24(6), 64–71.

[97] Alliance, C.H., White Paper on China Hydrogen and Fuel Cell Industry. China Hydrogen Alliance, Beijing. 2018.

[98] Meng, X., et al. Review of China's hydrogen industry policy and scientific and technological development hotspots in 2019. Science & Technology Review, 2020, 38(3), 172–183.

[99] Zhang, M. and Yang, X., The regulatory perspectives to China's emerging hydrogen economy: Characteristics, challenges, and solutions. Sustainability, 2022, 14(15), 9700.

[100] Zou, C., et al. Current Situation, Technological Progress, Challenge and Prospect of Hydrogen Energy Industry. Natural Gas industry, 2022, 42(4), 1–20.

[101] Dincer, I., Green methods for hydrogen production. International Journal of Hydrogen Energy, 2012, 37(2), 1954–1971.

[102] Yu, H., et al. Hydrogen Production by Water Electrolysis: Progress and Suggestions. Strategic study of CAE, 2021, 23(2), 146–152.

[103] IEA, *Global Hydrogen Review 2022*. 2022.

[104] Zhang, X., et al. Cost analysis on hydrogen production via water electrolysis. Modern Chemical Industry, 2021, 41(12), 7–11.

[105] Xu, S. and Yu, B. Current development and prospect of hydrogen energy technology in China. Journal of Beijing Institute of Technology (Social Sciences Edition), 2021, 23(6), 1–12.

[106] Dai, F. Study on Modeling of Catalyst and Direct-coupled Photovoltaic Power Generation System for Hydrogen Production from PEM Water Electrolytic. Hangzhou: Zhejiang University, 2020, n/a-n/a.

[107] Tian, J., et al. Development status and trend of green hydrogen energy technology. Distributed Energy, 2021, 6(2), 8–13.

[108] Liu, J. and Hou, T. Review and prospect of hydrogen energy storage technology and its application in electric power industry. Power & Energy, 2020, 41(2), 230–233.

[109] Ling, W., et al. Development strategy of hydrogen infrastructure industry in China. Strategic study of CAE, 2019, 21(3), 76–83.

Kandasamy Palanivelu* and Sathyanarayanan Sri Shalini

4 Sustainable circular economy in the Indian context: policies and best practices

Abstract: The resource security in India is one of the problems in continuing economic growth and raising standards of living. Due to rapid urbanization and high economic growth, the Indian urban population in particular is increasing, generating more waste. Circular economy (CE) in which resources, once accessed, continue to be used in perpetuity and India takes a lead in promoting CE because there is an urgent need for decoupling economic growth from resources. The concept of 5Rs – reduce, reuse, recycle, reproduce, and refurbish – is a clear pathway for the transition of the Indian economy to mere CE. Various sectors such as agriculture, automobile, construction and building, and electronics have huge potential in augmenting resource efficiency and resources with innovative applications. The CE policies and realization of the opportunities are discussed in various sectors in the Indian context.

4.1 Introduction

The phrase "circular economy" (CE) is an alternative to the traditional linear economy and is used these days in the context of initiatives aimed at driving resource efficiency (RE). The keyword here being "Circular," as these business models encourage a shift from linear value chains to circular value chains, thereby enabling more efficient and complete utilization of resources. A CE refers to a closed-loop system that aims to decouple economic growth and consumption of resources. The idea is to redefine value creation in a way that designs waste and pollutants out of the system, works toward regenerating natural resources, and ensures that products and materials are kept in use [1]. This concept of CE was first introduced by the European environmental economist [2], and suggested closed cycle of resource flow in the environment.

Acknowledgment: Dr. S. Sri Shalini gratefully acknowledges the financial support provided by the Department of Science and Technology (DST), Ministry of Science and Technology, Government of India, under Women Scientists Scheme-A (WOS-A) (grant no. SR/WOS-A/EA-37/2018).

*Corresponding author: Kandasamy Palanivelu,** Centre for Environmental Studies, Department of Civil Engineering, College of Engineering Guindy Campus, Anna University, Chennai 600025, Tamil Nadu, India; Centre for Climate Change and Disaster Management, Department of Civil Engineering, College of Engineering Guindy Campus, Anna University, Chennai 600025, Tamil Nadu, India
Sathyanarayanan Sri Shalini, Centre for Climate Change and Disaster Management, Department of Civil Engineering, College of Engineering Guindy Campus, Anna University, Chennai 600025, Tamil Nadu, India

https://doi.org/10.1515/9783110767179-004

Concerns over rapidly depleting vital resources and adverse impacts on natural environment have lately gained greater prominence, resulting in increasing focus on judicious use of resources globally through combination of conservation and efficiency measures and advocating transition toward CE.

4.1.1 India

India is considered now as the fastest growing economy in the world with unity in diversity comprising 28 states and 9 union territories. The population in India is about 1.4 billion as estimated in 2019 based on the most recent UN data and 2011 census data of India [3]. India is the second most populous country. According to the latest projections, the country could overtake China and become the world's most populous country by 2027. It is the seventh largest country in the world, having total area of 3,287,263 km^2 (1,269,219 sq mi) measuring 3,214 km (1,997 mi) from north to south and 2,933 km (1,822 mi) from east to west. A land frontier of 15,200 km (9,445 mi) and a coastline of 7,516.6 km (4,671 mi) exist in India.

India boasts about 18% of the world population while occupy only 2.4% of the world surface and thus claims only a small share of global resources. India, as one of the fastest growing economies with GDP at USD 2.6 trillion, has increased its material consumption to six times, from 1.18 billion tons in 1970 to 7 billion tons in 2015; however, this economic growth has been coupled with inherent cost on natural environment. India has seen an average growth of 6.74% in the last decade. The material consumption is projected to more than double by 2030, to provide for increasing population, rapid urbanization, and growing aspirations. The projected pace of economic development is going to put pressure on already stressed and limited resources and may lead to serious resource depletion and environment degradation affecting the economy, livelihoods, and the quality of life. Further, material use is also closely associated with the problem of increasing wastes, which when suitably processed could deliver valuable resources again.

4.1.2 Circular economy and sustainability

CE is a promising approach concerning sustainable operation management intended for sustainable use of resources [4, 5]. Krill et al. [6] suggested retaining the value of resources at the end of their life cycle to capitalize on the circularity process. Zhang et al. [7] proposed the methods, namely reuse, remanufacturing, and recycling, as the sustainable strategies of CE. Foundation [8] proposed the CE principle-oriented ReSOLVE framework that includes:

(i) Regenerate – based on the conversion of waste into a source of energy for different operations along the value chain.

(ii) Share – based on the sharing of resources to extend the life cycle via recovery operations from the economic point of view.

(iii) Optimize – based on a technology-centered strategy, which requires the organization to use digital manufacturing technology or structural reform to reduce waste in the operations system across the supply chains.

(iv) Loop – based on the restoration of the value of products via recovery operations.

(v) Virtualize – based on the service-focus strategy that allows virtual and dematerialized products.

(vi) Exchange – based on introducing advanced and renewable goods instead of old and nonrenewable goods.

(vii) For the above-given elements of the ReSOLVE framework, Urbinati et al. [9] pointed out that the CE business models should be based on diminishing the reliance on new materials. It should move to the renewable energy-based system to enhance embracing the sustainable operation practices.

India as a signatory to UN Sustainable Development Goals is committed to provide for sustained economic growth along with sustainable use of natural resources and safeguarding environment. In India, the percentage of people living in urban areas is 37.7% in 2015 as compared to 17.29% in 1951, and due to rapid urbanization, it has become important to develop and propagate effective waste management system by the government.

RE means to create more output as products/services using less inputs. It reduces waste, drives greater resource productivity, delivers a more competitive economy, addresses emerging resource security/scarcity issues, and very importantly helps reduce the associated environmental impacts. CE keeps resources in use as long as possible by extracting the maximum value, and recovering and regenerating products and materials at the end of each service life, so as to limit the extraction of natural resources to the maximum possible extent. RE offers benefits on multidimensional aspects of economic, social, and environmental well-being. Cost savings from reduced material use, resource security, reduced conflict and displacement, for example, from mining, employment opportunities in green jobs, reduced greenhouse gas emissions, pollution and ecological degradation among other benefits drive the cause of RE.

4.1.3 NITI Aayog

The Indian government is striving to transform from a linear economy to a CE [10]. Various RE- and CE-related strategies addressing issues in many industries helped to establish a broad framework for enhancing RE in the Indian economy such as the strategy for RE by the National Institution for Transforming India (NITI Aayog) [11]. Its key recommendations addressing all life cycle stages as well as crosscutting issues were threefold: first, the promotion of eco-labeling, standards, technology development, green public procurement, industrial clusters, and awareness; second, the regulation of economic instruments, viability gap funding, and policy reforms across life cycle stages; and third, institutional development, including capacity development, institutional setup and strengthening, database and indicators, and resource index as a part of an economic survey. Beyond this strategy of 2017, NITI Aayog, Ministry of Steel, Ministry of Mines, Ministry of Housing and Urban Affairs, and Ministry of Electronics and Information Technology furthermore released a strategy on RE for the steel sector [12], the aluminum sector [13], the C&D sector [14], and the EEE sector [15]. NITI Aayog has released the latest report on the status of RE and CE [16]. Many initiatives have been taken to transform from linear economy to CE, and to further accelerate the transition to a CE, 11 committees for 11 focus areas have been created and listed in Table 4.1 [10]. It is chaired by the concerned line ministries and delegates from the MoEFCC and NITI Aayog, field experts, academicians, and industry officials. The comprehensive action plans are formulated by committees for moving from a linear economy to a CE in those focus areas along with the implementation strategies. The 11 focus areas cover the end-of-life products or recyclable materials or wastes.

Table 4.1: Focus areas for transition from linear economy to circular economy.

No.	Focus areas	Concerned line ministry
1	Municipal solid waste and liquid waste	Ministry of Housing and Urban Affairs
2	Scrap metal (ferrous and nonferrous)	Ministry of Steel
3	Electronic waste	Ministry of Electronics and Information Technology
4	Lithium ion (Li-ion) batteries	NITI Aayog
5	Solar panels	Ministry of New and Renewable Energy (MNRE)
6	Gypsum	Department for Promotion of Industry and Internal Trade
7	Toxic and hazardous industrial waste	Department of Chemicals and Petrochemicals
8	Used oil waste	Ministry of Petroleum and Natural Gas
9	Agriculture waste	Ministry of Agriculture and Farmers' Welfare
10	Tire and rubber recycling	Department for Promotion of Industry and Internal Trade
11	End-of-life vehicles (ELVs)	Ministry of Road Transport and Highways

4.1.4 Sectoral standards

Some sectors follow their own strategy on RE to address their specific issues and have already accomplished standardization at different levels. According to the strategy on RE in aluminum sector [13], one issue identified has been scrap usage, which is diffused and not regulated through standards or end-use restrictions and is further characterized by heavy reliance on imports. As one of the goals, aluminum scrap standards (e.g., as implemented in the European Union and China) should be developed to improve the quality of recycled metal and reduce the processing cost. Relevant IS codes in the aluminum sector are the IS 733:1983: Wrought Aluminium and Aluminium Alloy Bars, Rods and Section; the IS 1253:1992: Aluminium for Use in Iron and Steel Manufacture; as well as ISO TC-226: Materials for Production of Primary Aluminium. In the steel sector, it was found that only minimum environmental standards are required and to be introduced for scrap metal facilities across the industry. Consequently, environmental concerns are high, especially where end-of-life vehicles and/or white goods are to be processed. Relevant IS in the steel sector are the ISO TC-17 SC-3 (P): steels for structural purposes; the ISO TC-17/SC 11 SCMTD 16 (P): steel castings; and the ISO TC-5 SC-1 (P): steel tubes. In the EEE sector, findings revealed that India has already developed guidelines and standards for new product development in the electronics sector; however, the standards for the use of secondary materials are not yet specified. Consequently, standards are needed for recycling to mitigate the environmental and health impacts of unsafe recycling in the informal sector. Some relevant ISs in the EEE sector are the IEC TC-111 (P): environmental standardization for electrical and electronic products and systems; the IEC TC-59A: performance of household and similar electrical appliances; and the IEC TC-104 (O): environmental conditions, classification, and methods of testing.

The draft National Resource Efficiency Policy (NREP) 2019 prepared by the Ministry of Environment, Forest, and Climate Change of Government of India [17] envisions a future with environmentally sustainable and equitable economic growth, resource security, healthy environment (air, water, and land), and restored ecosystems with rich ecology and biodiversity. The draft NREP is guided by the principles of (i) reduction in primary resource consumption to "sustainable" levels, in keeping with achieving the Sustainable Development Goals and staying within the planetary boundaries; (ii) creation of higher value with less material through resource-efficient and circular approaches; (iii) waste minimization; and (iv) material security, and creation of employment opportunities and business models beneficial to the cause of environment protection and restoration. The listed targets are:

– By 2030, domestic scrap will fulfill 50% of the total aluminum scrap requirement.
– Increase the recycling rate to 50% by 2025 and 90% by 2030.
– Increase the utilization rate of dross to 40% by 2025 and 80% by 2030.

4.2 Circular economy in India

CE is practiced in some form across various sectors in India like vehicle manufacturing, food and agriculture sector, cities and construction, steel, carbon capture and utilization (CCU), e-waste, solid and liquid waste management (SLWM), and plastics. The main points of these sectors are detailed below.

4.2.1 Circular economy in vehicle manufacturing

The Indian Government had launched the Smart City Mission in 2015 to motivate different transportation options in urban planning. The recommendations of the mission encourage cities to facilitate a variety of transportation alternatives: transit-oriented system, public transportation, and last-point para-transportation connection. The implementation of modern transport system could turn car ownership irrelevant in cities because of less owned car in India. As per the Helsinki Mobility Plan [18], an innovative coordinated transportation network that combines multiple types of public, semiprivate, and private transportation into an unified system which makes private cars obsolete in the region. The system can combine several mobility solutions on a single payment platform, involving public taxi, bus, car pool, and sharing motorcycles. Commuters can navigate on platform through a smartphone application. The application serves as a journey planner, whereby the commuter obtains suitable transport route just by entering start and end points. Vehicle-as-a-service, pay-and-use model practically eliminates the necessity of vehicle ownership, giving persons access to personalized transportation. Sharing platforms provide easy access to numerous cars as well as recent technology, hence increasing the utilization of cars. A mobile system can facilitate the sharing of trips with common destination, increasing vehicle utilization levels. This contributes to decreasing additional resources, yet raising resource utilization following one of principles of CE.

For example, Ola and Uber provide a mobile application for booking cab and auto-rickshaw. The organization is expanding an idea to share trips with person going in similar direction. Passengers can share the seats and also the cost of travel so that it can be affordable for many. Even lending of bikes, scooters, and automobiles for hourly basis is becoming popular these days. This makes idle time of resources (vehicles) efficient. In India also, vehicle-sharing services are getting popular with company such as Zoom car and Myles, providing hourly rate with self-driving facility. India is confronting air pollution problems in many cities because of an increase in vehicle numbers. Besides, to change the reliance on large number of transports running on fossil fuel, government has started looking for opportunities. Supporting electric vehicles is considered a practical approach to this problem. They have initiated National Electric Vehicle Plan 2020 in 2013. This program aims to increase the share of electric vehicle in the market by providing financial assistance.

4.2.2 Circular economy in food and agriculture

Resource and knowledge solutions are digitally accessible. Online networks provide platform to share equipment with small farmers. Many farmers take advantage of the shared knowledge through various digital tools. This has also led to the exchange of ideas with regard to seasonal crops, quality of seeds, and geographic location. A special supply chain for the food product has been designed on digital platform. Digital channels convey real-time information (such as price, size information, and position of demand) to farmers. Also, it links producers with consumers to directly benefit both the parties, cutting losses happening in the traditional marketing concept. ITC has launched a digital platform that enables farmers to look for information about government scheme and market location. This has facilitated the small farmers to get the actual price for their crop products. It has completely changed the supply chain of food materials.

4.2.3 Circular economy in construction and building

In India, the estimated consumption of construction materials is very high [19]. India generates about 150 million tons of construction and demolition (C&D) waste per year [20]. It fluctuates from 5% to 25% of the Municipal Solid Waste (MSW) generated [21]. It can only recycle 1% every year. The C&D waste is managed according to the strategies and framework by C&D Waste Management Rules, 2016. The best practices of the C&D waste management are detailed below: The Burari C&D waste management plant in North Delhi commenced in July 2009, and has a design capacity of C&D waste of 2,000 tons per day (TPD). The plant gets the average daily waste intake of 1,869 TPD (2019–2020). It is operated by the public–private mode, that is, urban local body (ULB) and private operator mode. The process uses both dry and wet wastes. The percentage of inert or reject going to landfill is less than 5%. The outputs from the process are paver blocks, cement bricks, kerbstone, tiles, recycled aggregates, and recycled concrete aggregates. The fee charged by ULB for collection and transportation is Rs. 239.44/MT with 5% annual increase paid to concessionaire by North Delhi Municipal Corporation (DMC). The processing fee of Rs. 205/MT is collected and retained by the concessionaire. There is no processing fee for waste of North DMC. The sale of products in 2019–2020 was 514,484 MT (private parties – Rs. 214,048 MT; ULBs – Rs. 112,493 MT; and other agencies – 187,943 MT). The other C&D waste processing plant in East Kidwai Nagar, New Delhi, set up the M/s Enzyme India Pvt Ltd in 2014, an eco-friendly process. It has a capacity of 150 TPD C&D waste recycling plant. The plant is operated by public–private partnership model with 100% buyback by NBCC. It produces 30,00 bricks per kerbstone per day which are utilized at the construction site.

Another application of C&D block usage is for the exterior wall construction in the additional office complex for Supreme Court of India. It is constructed by Central Public Works Department near Pragati Maidan, New Delhi. The properties of C&D block used were size: 400 × 200 × 100 mm; grade of concrete: M-10; and compressive strength: 10–15 N/mm^2. Instead of 3 kg of traditional bricks, the C&D block weigh 15 kg, but only 125 m^3/block was consumed as compared to 500 bricks/m^3 [21]. Gurugram city in Haryana generates C&D waste of about 1,200 TPD and to handle this waste a C&D waste plant was implemented in 2019 with a capacity of 300 TPD. The capacity of the plant further increased to 1,500 TPD. Approximately 3.5 lakh tons of C&D waste are processed from the unclaimed dumpsites [20].

Due to inadequate capacity of C&D waste processing facilities, the material value of C&D waste is lost into landfills, causing huge environmental and economic losses. C&D waste management helps to suppress dust generation, thus significantly reducing air pollution. Additionally, it conserves precious resources and minerals and helps in promoting the use of recycled products for construction and other infrastructural projects. Complete circularity in C&D waste management can be achieved by implementing a comprehensive strategy and action plan covering the life cycle of construction projects, including dismantling phase. The recommendations are:
(i)   reduction in virgin construction of raw material usage in different building projects;
(ii) extending tax rebates on recycled C&D products.

Calcium-rich industrial waste usage in construction materials, their properties and different applications through marble waste and flue gas desulfurization (FGD) gypsum. Large quantities of industrial wastes are stockpiled and haphazardly disposed in increasing amounts causing serious environmental concerns. The extensive use of marble and gypsum products is increasing in construction industry, and a limited amount of natural sources are available, which requires alternative sources of calcium-rich raw materials. Utilization of these industrial wastes leads to an increase in economic efficiency and it takes a positive step toward the conservation of natural materials, resource recovery, and protecting the environment. Approximately 22 million tons of marble waste and 20 million tons of FGD gypsum are expected to be generated by the year 2040 in India. The results of physicochemical analysis indicated that marble waste and FGD gypsum have the potential as raw materials for civil infrastructure [22].

4.2.4 Circular carbon economy

Building on this CE concept, a circular carbon economy (CCE) model focuses on adopting clean technologies to ensure energy market stability and, at the same time, guaranteeing inclusive growth that caters to the sustainable development goals. The basic idea is to manage carbon dioxide in the conventional thermal power stations. The four R concepts of CCE, namely reduce, reuse, recycle, and remove, are categories of mitigation options to be practiced, making carbon neutral in that order. "Reduce" includes all such mitigation options that reduce the quantum of CO_2 entering the system such as energy efficiency and other means including nuclear power. "Reuse" refers to capturing and using CO_2 as an input in processes that convert into useful feedstock for industry. "Recycle" refers to nature's carbon cycle wherein natural sinks such as plants and soil absorb carbon from the atmosphere and subsequently release it through decomposition and combustion. Thus, the bioenergy subsystem becomes carbon neutral, provided an equivalent biomass grows and replaces what was used as biofeedstock for bioenergy. Lastly, "remove" involves eliminating carbon from the system by converting the captured carbon into feedstock for reuse or removing it by storing it via chemical or geological processes [23].

4.2.4.1 India and CCUS

India's main carbon neutrality achievement is aimed in a natural way of shifting toward renewable energy and increasing the green cover. Still, India had identified carbon capture, utilization, and storage (CCUS) as a priority area in its second biennial update report that was submitted to the United Nations Framework Convention on Climate Change (UNFCCC). The country is an active participant in the Carbon Capture Innovation Challenge under Mission Innovation (MI). India launched a funding opportunity in 2018–2019 under MI for carbon capture (IC3), sustainable biofuels (IC4), and converting sunlight (IC5). The idea was to aid collaboration between Indian researchers and other MI member countries and a budget of USD17 million has been sanctioned for 47 projects (IC3 – 20, IC4 – 14, IC5 – 13) across the 3 themes. Additionally, as of July 2020, under the Indo-US Strategic Energy Partnership, CCUS technologies are among the common areas that have been jointly identified for collaboration.

The uptake of CCUS has been relatively slow in the country, primarily because of concerns regarding geological CO_2 storage, high costs, and uncertainties regarding such technologies. Despite the skepticism associated with CCUS, some independent companies have ventured into this field. For instance, in July 2019, the Oil and Natural Gas Corporation and Indian Oil Corporation Limited signed a Memorandum of Understanding (MoU) to jointly work toward reducing carbon emissions through the implementation of CCUS at the Koyali Refinery in Gujarat. Similarly, Dalmia Cement announced its plans to build a 500,000-ton carbon capture cement plant in Tamil Nadu. As of September 2019, it has signed an MoU with UK-based Carbon Clean Solutions Ltd. for technology and operational services for running the plant. A plant of this capacity is the first of its kind for this purpose in India. Dalmia Cement happens to be the first cement company in the world to have committed to becoming carbon negative by 2040. Additionally, a plant situated in the industrial port of Tuticorin captures CO_2 generated from its boiler and uses it to produce baking soda, which has a wide market base in industries such as glass making, detergents, and paper products. Small-scale capture and utilization plants for fertilizers are also in the works.

With regard to the increasing focus on "hard-to-abate" sectors, CCUS is envisioned to play a key role, particularly in the iron and steel industry. India is the second largest producer of steel and the country's steel industry is more energy and emissions intensive as compared to its global counterparts. Based on estimates by the IEA [24], new capacity additions in India over the next decade are projected to account for 40% of the steel-making capacity of the country, which would be in operation in 2050 (excluding a few early retirements). It therefore warrants the need for investment in near-zero emission technologies which need not be limited to the steel sector alone.

Tuticorin Alkali Chemicals & Fertilizers, a chemical and fertilizers company based in the South Indian state of Tamil Nadu, has made a global breakthrough in carbon capture technology, one that promises to prevent emissions of 60,000 tons of CO_2 annually. It also has the potential to push forward the circular agenda in India, which the Ellen MacArthur Foundation and United Nations Conference for Trade Development believe could put India on the path to regenerative and value-creating benefits.

According to the company, the plant is now close to achieving its zero-emissions goal, operating with almost no emissions seeping into air or water, thanks to a patented carbon-stripping technology from UK-based Carbon Clean Solutions. The technology employed at the Tuticorin plant converts captured carbon into soda ash, a base chemical used in glass manufacturing, paper production, and detergents.

The chemical strips CO_2 emissions from boiler chimneys through the form of a fine mist. As the chemical plant's coal-fired boiler releases flue gas, a spritz of Carbon Clean's new patented chemical removes the CO_2 molecules. To create soda ash, the captured CO_2 is mixed with rock salt and ammonia. While Tuticorin appears to be motivated by the financial benefits that the technology offers, Carbon Clean has suggested that it has the potential to capture between 5% and 10% of the world's coal emissions.

The groundbreaking project, which was launched in October 2016, is the first of its kind and is, interestingly, privately financed and this project is a game-changer. By capturing and crucially reusing CO_2 at just $30/ton, there is an opportunity to dramatically accelerate the uptake of CCU technology, with its many benefits, around the world. This is a project that does not rely on government funding or subsidies – it just makes a great business sense.

Unlike most carbon capture and storage projects, which typically bury CO_2 in underground rocks, the partnership between Tuticorin and Carbon Clean represents the first successful industrial-scale application of CCU, in which carbon is converted for reuse to turn a profit. Additionally, Carbon Clean claims that this new technology differs from current CO_2-derived chemical materials, in that the chemicals are less corrosive and require less energy and equipment handling.

India's leading Tata group has taken a strategic step in their journey toward decarbonization, commissioned a 5 TPD carbon capture plant at its Tata Steel Jamshedpur, making it the country's first steel company to adopt such a carbon capture technology that extracts CO_2 directly from the blast furnace gas. Tata Steel will reuse the captured CO_2 on site to promote CCE. This CCU facility uses amine-based technology and makes the captured carbon available for onsite reuse. The depleted CO_2 gas is sent back to the gas network with increased calorific value. This project has been executed with technological support from Carbon Clean, a global leader in low-cost CO_2 capture technology.

4.2.5 Circular economy in e-waste management

Electronic waste (e-waste) typically includes discarded computer monitors, mother-boards, mobile phones and chargers, compact disks, headphones, television sets, air conditioners, and refrigerators. E-waste is not just environmentally harmful but also has its social and economic implications as well, as this waste is handled by the informal sector without proper rights and conditions mainly in developing countries. India is also one of the largest e-waste producers, according to the e-waste global monitor. India generates about 3.23 million tons (in 2019) of electronic waste annually only from the organized sector, and including the unorganized sector, there is about 5 million tons generated in the country [25]. In terms of the resources embedded in the e-waste, glass waste is 37% and is the highest, metallic waste is 33%, and plastic waste is 30%. Metallic constituents of e-waste are iron 52%, zinc 3%, copper 18%, aluminum 12%, lead 3%, and others 12% [26]. E-waste is also increasing at the rate of 5% per year [19]. Mumbai (Maharashtra) generates high amounts of e-waste, followed by Delhi and Bengaluru.

In India, laws to manage e-waste have been in place since 2011, mandating that only authorized dismantlers and recyclers collect e-waste. The list of state-wise dismantlers/recyclers in India is given in Table 4.2 [27].

Table 4.2: List of state-wise dismantlers/recyclers in India as per the authorization issued by SPCBs/PCCs under E-Waste (Management) Rules, 2016 (as on April 29, 2022).

S. no.	State	Number of authorized dismantler/recycler	State-wise capacity (MTA)
1	Andhra Pradesh	8	32,122.5
2	Assam	1	120.0
3	Chhattisgarh	2	6,750
4	Delhi	2	120
5	Gujarat	33	84,301.92
6	Goa	1	103
7	Haryana	42	137,415.6
8	Himachal Pradesh	2	1,500
9	Jammu and Kashmir	3	705
10	Jharkhand	2	660
11	Karnataka	71	52,842
12	Kerala	1	1,200
13	Maharashtra	116	106,280.5
14	Madhya Pradesh	2	9,600
15	Orissa	5	5,690
16	Punjab	7	9,492
17	Rajasthan	24	83,604
18	Tamil Nadu	32	132,049

Table 4.2 (continued)

S. no.	State	Number of authorized dismantler/recycler	State-wise capacity (MTA)
19	Telangana	17	113,012
20	Uttar Pradesh	89	494,042.7
21	Uttarakhand	6	153,125
22	West Bengal	4	1,950
Total		472	1,426,685.22

Manufacturers, dealers, refurbishers, and producer responsibility organization (PRO) were brought under the ambit of the E-Waste (Management) Rules, 2016. The National Resources Policy also envisages a strong role for producers in the context of recovering secondary resources from e-waste.

E-Parisaraa Pvt Ltd was established in the year 2004 in Bangalore, India, started operations since September 2005 which is India's first government-approved Electronic Waste recycling company focusing on recycling services. The company recycle end of life, obsolete, discarded and custom de-bonded electronics and electrical equipment like computers, CPUs, servers, printers, fax machines, copiers, mother boards, printed circuit boards, CDs, floppies, tapes, cartridges, telephones, cell phones, telecom equipment, TVs, audio and video, dry cells, lithium batteries, fluorescent and CFL lamps, household microwave, washing machines, industrial and household, medical, military, and space electronics in environmental-friendly way and convert waste electronic and electrical equipment into raw materials like metals, plastics, and glass. Their services include witness crushing, assured destruction, certificate of destruction, downstream material accountability, and precious metal recovery. There is recovery of apparent gold of printed circuit boards (PCBs) and components, and reuse of the same for electroplating of temple items, watch and pen parts, imitation jewelry, and so on [28].

Cerebra Integrated Technologies Limited is also a Bengaluru-based company [29]. The company is engaged in the business of e-waste recycling, refining and refurbishment, electronic manufacturing services, and information technology infrastructure management with a philosophy called "Circularity" by not only remanufacturing but also by responsibly recycling the products that have reached the end of their life. The e-waste plant capacity of recycling: 97 K MT and refurbishing 240 K units

Jamshedpur (Jharkhand) generated about 230 tons of e-waste [20]. It has a centralized e-waste collection center and five decentralized e-waste collection centers. The public–private partnership of e-waste management is adopted between the Jamshedpur utility service company and Hulladek Recycling Pvt Ltd, the PRO for e-waste management in Jamshedpur. Hence, the ULB has no cost for e-waste processing. The collected e-waste is sent to the e-waste management center having a capacity of 35 ton, followed by transportation to the Kolkata warehouse of Hulladek monthly once. From the warehouse, the e-waste is sent to six recyclers.

Attero Recycling, a premium e-waste company, provides end-to-end recycling solutions to electronic scrap such as television sets, computer monitors, printers, scanners, keyboards, mice, cables, circuit boards, lamps, calculators, phones, answering machines, and DVDs [30]. The unique recycling process is a mix of mechanical and hydrometallurgical technologies that can extract 98% of metals from e-waste, ensuring low carbon dioxide emissions. The company adopts methods like PCB recycling that involves recovery of copper, silver, and gold from PCBs, tin from tin/lead solder dross, catalytic converters to extract platinum, palladium, and rare earths like neodymium for magnets. The company also claims to be the only one to recycle Li-ion batteries irrespective of size, shape, and chemistry in an environmentally friendly manner. Currently, at their Roorkee facility, they recycle close to a thousand tons of Li-ion batteries every year. From a CFL bulb to industrial equipment, the company recycles more than 20 types of e-wastes in its facility that has capacity to treat 144,000 tons of waste annually.

Due to strict enforcement, now, lot of e-waste recyclers are growing in India, and the country has immense potential in augmenting e-waste recycling. There are some forward movements in this direction as evident from the list of authorized recyclers (Table 4.2) approved by CPCB/SPCB, which reveals that there are 472 authorized recyclers with a total capacity of 1,426,685.22 MTA in India as on April 29, 2022. Still, lots of ground must be covered through skill development, building human capital, and introduction of clean technology while adopting adequate safety measures in the country's informal sector. Since India is highly deficient in precious mineral resources, there is need for a well-designed, robust, and regulated e-waste recovery regime which would generate jobs as well as wealth.

4.2.6 Circular economy in steel industry

India is the second largest producer of steel in the world after China, with a production of 106.5 MT in 2018, which is almost 6% of global crude steel production [31]. Iron and steel industry is a high-energy and capital-intensive industry. This sector contributes to about 2% of country's GDP.

Jindal Steel and Power Limited is an Indian Private Limited company which contributes to the major production of steel components, power, and building structure. It is one of diverse parts of O.P Jindal group. This industry has manufactured the world's longest 121 m rails, flange plate, twist iron, and TMT rebars. It has the world's biggest coal-based sponge iron plant, which has the functional capacity of 3.25 MT per annum (MTPA). The company has the maximum ability to produce throughput of 6.75 MTPA of steel and plans to expand holistically, significantly contributing to India's sustainable growth.

Tata Steel has also made progress in its steel recycling business initiative, which is a definitive step toward sustainable steel production. The company has set up its first steel recycling plant at Rohtak in Haryana, which will enable lower carbon emissions, resource consumption, and energy utilization.

4.2.7 Circular economy in textile industry

The textile industries are nowadays forced to go for zero liquid discharge to minimize the pollution from their industrial effluents. The effluent stream bearing high salt, that is, the spent dye bath is segregated and treated with recovery of salt. This effluent stream has low volume and can be treated using a chemical treatment followed by multieffect evaporation and crystallization. Glauber's salt (sodium sulfate decahydrate $Na_2SO_4 \cdot 10H_2O$)-based dyeing enables recovery of the salt at about 12 °C. The other effluent stream, that is, wash water can be treated separately by primary/secondary treatment methods followed by RO system to recover water.

A case study of treatment scheme based on the textile industrial effluent recovery of water and Glauber's salt is presented in the CPCB report [32] installed by an industry engaged in the processing of hosiery fabrics and yarns with an average production of 5 tons of hosiery cloths and yarns per day. During processing, that is, scouring, bleaching, and dyeing, the contaminated wastewater is generated which is about 500 KLD (kiloliters per day). The treatment of wastewater in the scheme is primary treatment for dye color removal, reverse osmosis membrane filtration, multiple effect evaporators, and crystallizer. These measures enable the industry to recover water and Glauber's salt (sodium sulfate) for recycle in the production process. The total water recovery in the system is 463.75 KLD which is 92.75% of total waste water discharge of industry. Besides the water recovery, the system facilitates salt recovery which is estimated to be about 2,000 kg/day. For the 500-KLD plant, the industry recovers water and sodium decahydrosulfate, which gives returns of Rs. 76.54 lakhs per year.

Though nanofiltration is reported to allow maximum passage of the common salt with no color in the permeate. As such the permeate can be directly recycled back to dye bath so that fresh addition of salt can be reduced drastically. This has to take momentum. Currently, the NaCl salt used is mainly managed by evaporating the salt solution from RO reject and storing it. Ways should be found to recycle or reuse the NaCl too. Pure water obtained as RO permeate is used again in textile industries, thus minimizing the water requirements.

4.2.8 Circular economy in tannery industry

Tanneries doing chrome tanning process have installed chrome recovery units in their premises [33]. These tanneries segregate the chrome liquor and collect in a tank for precipitation of the chrome by adding magnesium oxide alkali solution. The precipitated chrome slurry is added with sulfuric acid to regenerate chrome and filled in carboys. Thus, the chrome is recovered and mixed with fresh basic chromium sulfate for reuse in the tanning process again.

4.2.8.1 Chromium recovery

a. Recovery of chromium for reuse in the tannery: All the chrome tanning units will install chrome recovery plant either on an individual basis or on a collective basis in the form of chrome recovery plant and use the recovered chrome in the tanning process. Re-solution of the chromium precipitated from the tanning float, using sulfuric acid for use as a partial substitute for fresh chromium salts. Applicability is restricted by the need to produce leather properties that meet customers' specification, in particular, related to dyeing (reduced fastness and less brightness of colors) and fogging.
b. Recovery of chromium for reuse in another industry: Use of the chromium sludge as a raw material by another industry. This applies only where an industrial user for the recovered waste can be found.

For chrome recovery and reuse, the operating cost for 5 m^3 of pickling liquor/day is about Rs. 2,000, while the value of Cr recovered is about Rs. 1,250. This technology is suitable for all kinds of skins and hides [34].

4.2.9 Circular economy in municipal solid waste and liquid waste management

The Swachh Bharat Mission (SBM) was started on October 2, 2014, on the occasion of Mahatma Gandhi's birth anniversary [35]. The main aim of this mission was to clean the lanes and roads, construct public toilets for eradicating open defecation, and create awareness of solid waste management and develop scientific management for it. The several schemes implemented are: (i) SBM for Urban Areas; (ii) SBM (Gramin); (iii) Swachh Vidyalaya Abhiyan; and (iv) Rashtriya Swachhata Kosh. Others are Swachh Sarvekshan, an annual cleanliness survey for the MSW services; Star Rating of Garbage-Free Cities; Swachhata Hi Seva for the single-use plastics; and so on. The total waste generated in India is about 140,557 MT/day from 89,061 wards with a total waste processing capacity of 70% [36]. The state-wise distribution of the waste generation and its processing capacity is shown in Figure 4.1. Due to the SBM activities, the waste processing capacity in India is increased from 18% to 70%. The Jan Andolan approach played a major role, where a change in behavior toward the Swachhata in the urban areas was noticed. The SBM is aiming at SLWM in the country. The SBM (G) Phase-II initiatives aimed for improved practices in SLWM. Following are some of the case studies for the SLWM in different states:

4.2.9.1 Solid waste management

Indian MSW is approximately 40–60% compostable, 30–50% inert waste, and 10–30% recyclable [37].

The inert content of the solid waste is decreasing and the content of recyclables such as paper and plastic are increasing over the period [38]. The generated various components of solid wastes are to be managed according to the Solid Waste Management Rules, 2016. The legal framework and institutional frameworks of the Government of India on solid waste management are detailed in [39]. Solid waste management can be further categorized as (a) biodegradable waste management, (b) Galvanizing Organic Bio-Agro Resources Dhan (GOBARdhan), and (c) sanitary waste management. The major issues with plastic waste and its management are discussed in the following sections of plastic waste management.

4.2.9.1.1 Biodegradable waste management

India generates about 400 million metric tons (MMT) of biodegradable waste every day [20], which mainly consists of fruits, vegetables, meat/fish and poultry, flowers, and park and green trimmings. The biodegradable waste is used to produce compost, biogas, and electricity. The total waste to compost generation was around 3,182,435 MT and WtE was 61 MW in 2020 [35]. Biomethanation can be profitable than traditional composting (for 3 lakhs and above population), only if there is proper market for end-product usage [21].

India's first largest compost plant capacity of 500 tons/day was operated by Excel industries Ltd., in Mumbai in 1992. The different composting technologies available in India and their distribution in different states along with their operational details are detailed elsewhere [39]. Mysuru, Karnataka, is having zero-waste management plants in every zone [20]. It receives biodegradable waste from five wards and is scientifically processed to compost and then sold to nearby farmers and horticulture department. Mysuru has a centralized compost plant with the capacity of 200 TPD, decentralized waste management plant of 35 TPD, dry waste collection center of 43 TPD, and centralized landfill of 90 TPD. Vengurla, Maharashtra, processes 100% of its organic waste generated. The city entitles to a no-landfill city. They collect the waste in a two-bin system, where a green color bag for biodegradable waste collection and a blue color bag for

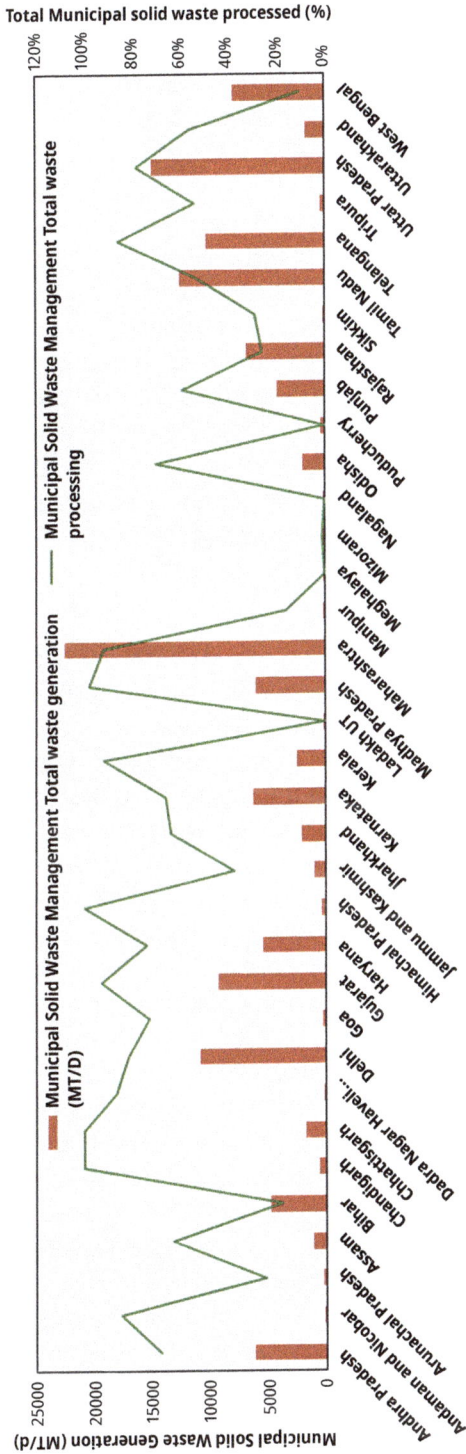

Figure 4.1: Municipal solid waste generation and processing capacity in different states of India (as of 2021).

nonbiodegradable waste. Other wastes such as sanitary and domestic hazardous wastes are collected separately. The kitchen waste is treated using vermicomposting, and fish/meat/fibrous wastes are treated using biomethanation. The city produces 2.7 TPD of biodegradable waste, where 2.5 TPD is treated at the central processing facility and 0.208 TPD in the decentralized facility. The compost quantity of 54 tons is produced and 4.7 tons were sold at Rs. 10/kg to farmers annually.

Similarly, Bobbili in Andhra Pradesh is one of the top 10 municipalities in India for waste processing. The city uses the Internet of things (IoT) and information and communication technology tools for real-time monitoring systems for waste collection and monitoring. It is processing all its organic waste through vermicomposting and/or windrow composting and biogas plants, and generates revenue from processing and recycling. Bobbili produces 120 tons of windrow compost every 2 months and 60 tons of vermicompost annually which is sold at Rs. 10/kg. The city spends Rs. 1.53 crores for MSW management and gains revenue of 1.57 crores from it. Tamil Nadu has 66,310 numbers of Thooimai Kaavalars for SWM activities [40]. Each gram panchayat has compost pits (two), pit for nonrecyclable/inert waste and vermicompost units.

4.2.9.1.2 GOBARdhan

India is the world's largest milk producer and it generates more than 1,500 MMT of cattle dung/year, where 992 MMT is recoverable and used for biomethanation [20]. The Galvanizing Organic Bio-Agro Resources Dhan (GOBARdhan) scheme is implemented in February 2018 for the safe handling of cattle waste and biodegradable or agrowastes and their conversion to biogas for domestic/commercial use. The bioslurry from the above process is used as fertilizer. The advantages of this scheme are income generation, compost fertilizer, safe power generation, better living standards, infrastructure development, and Green India and Clean India. The biogas can be operated in four models such as individual household model, community model, cluster model, or compressed biogas model through cooperatives/entrepreneurs/gaushalas. It aimed to cover more than 700 projects across the country in 2023–2024, especially one project per district for biowaste management [41].

The various waste to energy plants in India are detailed elsewhere [39]. The different waste to energy technologies in India are incineration (thermal process), pyrolysis and gasification (thermochemical process), biomethanation (biochemical process), and refuse-derived fuel (RDF, mechanical and thermal process). The incineration plants are spread in the states of Delhi, Telangana, Tamil Nadu, Madhya Pradesh, and Himachal Pradesh. Four biomethanation plants are in the states of Bengaluru, Karnataka (Nobel Exchange), with an installed capacity of 250 TPD and output of manure 25 TPD; Pune, Maharashtra (Nobel Exchange), with a capacity of 300 TPD and output of Bio-CNG 4TPD and manure 7.5 TPD; Solapur, Maharashtra (Organic Recyclers), with a capacity of 400 TPD and output of electricity 3 MW and manure 60 TPD; and lastly, Chennai, Tamil Nadu (Ramky), with a capacity of 30 TPD and output of electricity 0.26 MW and manure 3 TPD.

The Banas dairy operated in Banaskantha district of Gujarat, Rajasthan, and Uttar Pradesh has established a biogas plant at Dama Semen station for the cow dung waste [42]. The digester capacity is 3,000 m^3 and raw biogas is collected in double membrane balloons. The BioCNG capacity of 800 kg/day is produced, which can fill 100 vehicles. The solid fertilizer of 8 tons produced per day is further processed through vermicomposting for fertilizer production, and liquid of 70,000 L/day was sold to farmers to use as manure and retain moisture content in the soil. The project has revenue generation of Rs. 60,407,500/year, with the operating cost of Rs. 46,811,000 and yields a net profit of Rs. 13,596,500 with a payback of 5.4 years. The other advantages of the biogas plant are in reducing 800 kg/day of CH_4 emissions equivalent to 20 tons of CO_2/day and also 2.87 tons of CO_2/day when 800 kg CBG replaces petrol as fuel. The farmers are paid Rs. 1 per kg of cow dung. The payment of cow dung is paid to farmers along with the milk payment on every 15 days. These advantages enable 50 more plants to be built in the district. National dairy development board has implemented an efficient manure value chain [43]. The dung generation was about 1,653 MMT in 2018–2019. The recoverable dung of 992 MMT can generate biogas equivalent to 50% of country's LPG consumption/year with values up to Rs. 444 Bn and bioslurry equivalent to 44% of country's NPK requirement with values up to Rs. 425 Bn. The case study of the Zakariyapura pilot in Anand showed additional income of Rs. 1/L of slurry sold to farmers, with potential monthly earning of Rs. 3,000 with a payback of biogas plant in a year. The slurry is used as a biofertilizer and the payback on the slurry processing plant is about 4 years. The liquid and solid biofertilizers are sold at Rs. 15/L and Rs. 20/kg, respectively.

4.2.9.1.3 Sanitary waste management

Menstrual waste is considered as sanitary waste under the Solid Waste Management Rules, 2016. Menstrual waste consists of menstrual absorbents soiled with blood and human tissue remnants, and menstrual hygiene products include cloth, sanitary napkins, and other materials for absorbing the menstrual blood. The different disposal options for the menstrual absorbents are deep burial, disposal into pit latrine, composting, incineration, pit burning, and so on. The best practices are deep burial at least a meter deep; composting it with the leaves, wet biomass, and dung slurry; incineration in the waste pit/drum having a shaft at heights so as to release the emissions; collecting the used cloths and napkins from schools and other places; and biosanitizers [41]. Healthcare facilities (HCFs) using common biomedical waste treatment facility are 131,837, and approximately 21,870 HCFs have their own treatment facilities on-site [44]. Karad, Maharashtra, demonstrated a 100% of sanitary waste collection and treatment in the CBWTF [20]. Pune, Maharashtra, conducted a Red Dot campaign, where the sanitary waste is wrapped and a red dot is marked for the proper handling and disposal of sanitary waste. The city also aimed for the value-added products from sanitary wastes.

4.2.9.2 Liquid waste management

India generates about 61,754 MLD of sewage per annum of which treated capacity is only 37% and the remaining 63% is discharged into water bodies without treatment [45]. The liquid waste management can be subdivided into (a) graywater management (GWM) and (b) fecal sludge management (FSM).

4.2.9.2.1 Graywater management

Graywater is the wastewater generated from the domestic activities of bathing, washing, and laundry. GWM is the treatment and reuse of domestic wastewater that is not contaminated with urine or fecal matter. According to GWM in rural India, Department of Drinking Water and Sanitation, Ministry of Jal Shakthi, India generates about 31,000 L of graywater every day in rural areas [41]. The different technologies available for GWM at household level are soak pits, leach pits, magic pits, and kitchen garden; and for community level are community soak pit, community leach pit, community kitchen garden, waste stabilization ponds, phytorid technology, anaerobic baffled reactor, duckweed pond system, constructed wetland, and soil biotechnology. A case study from Chand Samand gram panchayat, Karnal, Haryana, showed that it has wastewater stabilization ponds with three pond systems built under Mahatma Gandhi National Rural Employment Guarantee Scheme (MGNREGS) for treating the graywater and use it for the gardening and irrigation purposes, thus reducing the freshwater requirement. Wadala, North Solapur, Maharashtra, has 938 soak pits for GWM, which also increased the groundwater recharge in the area.

Tamil Nadu has indigenous technical-type design models for GWM and installed in Namakkal district in 2016–2017 [40]. The success of the model is expanded to the whole state. The guidelines for the individual and community soak pit designs were issued by the Government of Tamil Nadu MGNREGS during 2017–2018. As per 2020–2021, the total individual household soak pits were 700,000, community soak pits were 100,000, community soak pits with horizontal filter type were 300, and community soak pits with vertical filter type were 1,200. The SHGs in Madurai, Tamil Nadu, pilot the women-only construction model for the construction of household soak pits. A total of 2,070 village panchayats were chosen in 2020–2021 for decentralized GWM facilities. The Government of Tamil Nadu has released a handbook on *WASH – Water, Sanitation and Hygiene* in 2020, covering the aspects of sanitation, and solid, liquid, and plastic waste management.

4.2.9.2.2 Fecal sludge management (FSM)

Fecal sludge is in a slurry or semisolid form, which results from the collection, storage, or treatment of combinations of excreta and black water, with or without gray water. "Septage" is the liquid and solid waste that is pumped from a septic tank, cesspool, or onsite treatment facility after it has accumulated over a period of time [46].

The safe containment, handling, and disposal of black water or fecal matter from the household and community toilets are FSM. The country has a national policy on fecal sludge and septage management (FSSM), 2017, a supporting document for the states containing the information content, directions, and strategies for the proper implementation of FSSM across urban India.

Accordingly, some of the technological options available for the FSM are septic tank with drain field, septic tank with soak pit, septic tank with small bore system, septic tank with baffles, twin pit pour flush latrine, and fecal sludge treatment plant (FSTP). The regulatory guidelines and frameworks for fecal sludge and septage management in different states are given in Table 4.3 along with the status of FSTP units [47, 48].

Table 4.3: Regulatory guidelines and frameworks for FSSM in different states.

S. no.	States	FSSM guidelines	Status of FSTP units
1.	Andhra Pradesh	– Fecal Sludge and Septage Management: Policy and Operative Guidelines for Urban Local Bodies in Andhra Pradesh – Andhra Pradesh Government Order 134, March 2017	Cotreatment in existing/proposed STPs, 28; independent FSTPs, 77 covering all towns; total investment, Rs. 259 crores
2.	Maharashtra	– Guidelines for Septage Management, 2016 – Government Resolution to Move Beyond ODF to ODF +/++, 2017 – Maharashtra State FSSM Strategy – Government Resolution on Cotreatment of Fecal Waste at STPs, 2018 – Government Resolution on Setting Up Independent FSTPs at Scale, 2019	Cotreatment in existing STPs, 69; independent FSTPs (colocated at SWM plant), 327 covering all towns; total investment, Rs. 45 crores
3.	Odisha	– Odisha Urban Sanitation Strategy – Odisha Urban Sanitation Policy (2016) and ULB's Regulation (2018)	Cotreatment in existing STPs, 2; independent FSTPs, 97, all cities covered; total investment, Rs. 298 crores
4.	Rajasthan	– Draft Policy on FSSM, 2017 – State FSSM Guidelines for Urban Rajasthan, 2018	Rajasthan is preparing detailed project reports of more than 100 towns for FSSM implementation
5.	Tamil Nadu	– Tamil Nadu Septage Management Operative Guidelines, 2014	Cotreatment in existing STPs, 50; independent FSTPs, 59, population covered 2.5 crores (75% urban pop; 600 towns); total investment, Rs. 200 crores

Table 4.3 (continued)

S. no.	States	FSSM guidelines	Status of FSTP units
6.	Telangana	– The 2018 State Faecal Sludge and Septage Management (FSSM) Policy	Cotreatment in existing STPs; in Hyderabad (12); independent FSTPs, 71 (PPP-HAM) + 70 (EPC), covering all towns; total investment Rs. 250+ Cr
7.	Uttar Pradesh	– Guidelines for FSSM in Uttar Pradesh, 2018 – Draft State FSSM Policy, 2019	Uttar Pradesh has also undertaken cotreatment and FSTP implementation in 50+ towns

Some of the leading practices in emptying and conveyance, treatment and operations, integrated models for transport and treatment, and reuse and resource recovery are detailed below. The sanitation credit model in Jalna, Maharashtra, along with the building of individual toilets is a working practice and includes in other places of the country such as mission for elimination of poverty in municipal areas in Telangana and Andhra Pradesh; Kudumbashree in Kerala; involvement of women self-help groups (SHGs) in Telangana; innovative private sector model by Saraplast in Pune, Maharashtra; common septic tanks in Bhubaneswar, Odisha; and recently, inclusion of standard septic tank design and inspection under building rules in Tamil Nadu in 2019. Some states have adopted different state models for the emptying and conveyance of fecal sludge: full private model in Andhra Pradesh, Telangana, and Tamil Nadu; full government model in small cities of Maharashtra; PPP annuity model in Wai and Sinnar of Maharashtra, Leh, and UT of Ladakh; and government owned and leased to private players or SHG in Odisha. The mechanized desludging practice is intensified in Odisha. Digital technologies are used in FSSM in Maharashtra, Telangana, Odisha, and other states.

The Devanahalli FSTP in Karnataka has a capacity of 6 KLD, and completely works on gravity flow. It is the first FSTP under unplanted drying bed technology known as Devanahalli model. The produced soil conditioner is sold to farmers, which contributes to 29% of revenue in FSSM. The Government of Tamil Nadu implemented a cotreatment model in 2018 to allow cotreatment of fecal sludge at the existing STP. The colocation of FSTPs with solid waste management plants was observed in Maharashtra, Odisha, and Tamil Nadu. FSTP in the Resource Recovery Park at Periyanaickenpalayam town panchayat, Coimbatore, has a capacity of 25 KLD. It has been colocated within the SWM plant which also facilitates co-composting. A total of 311 FSTPs in Maharashtra were sanctioned in 2019; 120 FSTPs are constructed and 100 FSTPs are under progress, and are colocated within the SWM plant. The FSTP in Dhenkanal, Odisha, also employs co-composting technology.

4.2.9.3 Plastic waste management (PWM)

The current waste generation in India is approximately 1.45 lakh MT of solid waste/day, which has 35% of its dry waste [21]. The various dry wastes are plastic, paper and cardboard, glass and ceramic, metals, textiles, tires and rubbers, and so on. Among them, the plastic waste generation was about 3,360,043 tons annually [20]. The Plastic Waste Management Rules, 2016, provide the framework for efficient management of plastic waste and several amendments to the rules in 2018; August 2021; the second amendment in September 2021 and recently launched the Plastic Waste Management Amendment rules, 2022, in February 2022 [49]. It prohibits the manufacture, import, stocking, distribution, sale, and single use of plastic carry bags made of virgin or recycled plastic, which shall not be less than 75 μm in thickness till December 31, 2022, and after that 120 μm. The guidelines on Extended Producers Responsibility on plastic packaging are given in PWMR, 2022. It is a framework to strengthen the CE of plastic waste and the use of alternatives for sustainable plastic packaging. Single-used plastic items prohibited from July 2022 are:
a) plastic sticks of earbuds, balloons, candy, and ice cream;
b) cutlery items of plates, cups, glasses, spoons, forks, knives, trays;
c) packaging/wrapping films for sweet box, invitation cards, cigarette packets; and
d) polyvinyl chloride (PVC) banners less than 100 μm and polystyrene (thermocol) for decoration.

The generated plastics commonly end up in oceans, making up more than 80% of all marine debris [50]. Plastic waste accumulation can be avoided by recycling it as much as possible. Seven categories of plastics based on the recyclability are polyethylene terephthalate, high-density polyethylene, PVC, low-density polyethylene, polypropylene, polystyrene resins, and multimaterials like acrylonitrile butadiene styrene, polyphenylene oxide, polycarbonate, and polybutylene terephthalate. The plastic packaging that cannot be recycled or used as an alternate source of energy will be phased out. The multilayered plastic packaging can be co-processed and used as an alternate source of energy in waste to energy, cement kiln, road construction, and so on. The ULBs aimed to use plastic waste for road construction, energy recovery, waste to oil, and so on.

The Indian government has brought out a manual on plastic waste management under the SBM initiative for managing plastic waste, especially in rural parts of India [51]. This manual is an advisory for the villagers to know about the impacts of plastic waste, creating systems for source segregation and storage facilities for the plastic wastes, transportation and setting up of plastic waste management unit, highlighting the roles of different stakeholders and financial management for plastic waste management. The key principles of plastic waste management are the 4Rs of refuse, reduce, reuse, and recycle, where refuse, reduce, and reuse are the duties of households, and the door–door collection and segregation are the duties of gram panchayats. The recyclable plastics are handed over to scrap dealers for recycling. The nonrecyclable plastics shredded or separated combustible fractions recovered at cement industries or road construction. Twenty Indian states and union territories have completely banned plastics and five states banned partially in April 2018 [41]. Very recently, the central government also banned throughout India. The technologies mainly available for plastic waste management are recycling, incineration, and landfilling. It was showed that the recycling of 1 ton of plastic waste can save up to 5,774 kWh of energy. The plastic can be controlled by carrying their own cloth/paper bags and avoiding products made of plastic packaging can reduce the generation of plastics; can reuse plastic jars, bottles/containers for storage; recycling the plastic by remolding it into storage bottles, toys, buckets, and other usable items and recover the plastics as used plastic waste as a fuel substitute as RDF. Some of the case studies for plastic waste management in India are detailed below. The plastic smart Aryad Gram Panchayat in the Alappuzha district of Kerala has set to introduce recycled value-added products and use shredded plastics in road construction. The green force volunteers collect plastic wastes from 7,000 houses and 500 mercantile institutions and transport them to the plastic shredding unit. The shredded plastics are sold to the clean Kerala company and other private agencies. The user fee of Rs. 30 is collected from households and Rs. 100 from each institution every month. The green force volunteer is paid Rs. 6,000 as monthly emolument. Sikkim is the first state in India to ban on disposable plastic bags. It also controlled single-use plastic bottles in 1998. Later in 2016, it restricted plastic water bottle usage in government offices and at its events. The state also banned styrofoam and thermocol disposable plates and cutlery. Following Sikkim, 17 states and union territories banned plastic. The Sirmaur district in Himachal Pradesh uses plastic waste for making polybenches and polytoilets, and in road construction.

Plastic wastes in road construction: In Tamil Nadu, road length of 1,000 m in different stretches was laid using plastic waste as an additive in the bituminous mix, which worked well. It was supported by the scheme known as "1,000 km Plastic Road" [41]. Similarly, in Bengaluru, Karnataka, more than 2,000 km of several road stretches used plastic waste and performed well. Twenty-three tons of plastic waste was used for laying 10 km length roads on Bangalore University roads. Plastic waste usage in road construction was also implemented in Delhi road stretches of 50 km. It was shown to perform well. In Shimla, Himachal Pradesh, pilot testing gave a good performance of plastic waste roads. Later, each kilometer of road laid uses 1 ton of plastic waste, which overall reduces the cost of Rs. 35,000–40,000 due to less bitumen usage.

Co-processing of plastic wastes in cement kilns: The plastic wastes can be used as alternate fuels in the cement kilns. In Kumbakonam, Tamil Nadu, the plastic waste from legacy plastic from Karikulam dumpsite and new plastic waste are converted into RDF for cement industries and the remaining wastes are recycled [20]. Sixteen tons per day of plastic waste are nonrecyclable and 2 tons per day of recyclable plastics. Initially in 2015, the shredded plastic is sold to contractors at the rate of Rs. 15/kg and 16 km of new roads were laid down. Later, Kumbakonam municipality linked with Zigma Global Environ Solutions Pvt Ltd, to reclaim the Karikulam dumpsite through biomining. Kumbakonam is the Tamil Nadu's first successful biomining project for MSW dumpsite reclamation. The waste processed and restored through biomining is 100,000 m^3 and area restored is 5 acres, where the resource recovery facility is set up for segregating 14 categories of wastes. The recovered nonbiodegradable waste was sent to cement and agarbatti factories and recyclers. The Zigma Global Environ Solutions company pay Rs. 500 to industries (Dalmia and Ultratech cement factories) for processing plastic waste. The other nonbiodegradable waste such as coconut shells and liquor bottles were used in a productive way.

Options for plastic waste recovery: The plastic waste is converted into tiles and traffic cones [51]. In Ahmedabad municipal corporation, plastic waste is converted into irrigation pipes and tarpaulins. A 50 MT plant for shredding plastics and paper was installed.

4.2.9.4 Material processing from municipal solid waste

Bhopal, Madhya Pradesh, is processing all of its waste due to 100% segregation of waste at the source [20]. The city has six material recovery facilities (MRFs) for processing the nonbiodegradable wastes of 565 TPD and five biodegradable waste processing plants including three windrow composting of 410 TPD and two biogas plants of 105 TPD. The rag pickers play a major role in the collection of recyclables from nonbiodegradable waste. The dry waste is categorized into more than 10 categories and sent to recyclers. The RDFs are sent for energy recovery.

Dhenkanal district in Odisha also achieved 100% material processing, especially with help from SHGs. This is the first municipal corporation in Odisha to set up an MRF. Decentralized waste management is compulsory in 114 ULBs in the state. Dhenkanal has five microcompost centers and three MRFs. The recyclables and nonrecyclables are separated out from nonbiodegradable wastes. The recyclables are sold and the revenue is used for managing the MRFs and paid to SHGs. The Ecokart Technology Pvt Ltd collects 150 tons of nonrecyclables monthly and sent them to Bargarh Cement Plant as RDF. The city also makes PVC paver blocks from nonrecyclable plastic waste. Jamshedpur, Jharkhand, has Dry Waste Collection Centers for material recovery from nonbiodegradable waste. The city has a decentralized approach and innovative processing of plastic waste and e-waste recovery. The rag-pickers involvement is the success of the waste management. In Surat, adoption of a multipronged approach of achieving 100% source segregation and channelization of recyclables and RDFs to achieve a high material processing efficiency has resulted in substantial reduction of the waste received in the city's landfill.

4.2.9.5 Recommendations for the circularity in waste management

Various recommendations from the Government of India for the circularity in waste management [21] are summarized in Table 4.4.

Table 4.4: Recommendations for the circularity in waste management.

S. no.	Waste category	Recommendations
1.	Dry waste	i) A comprehensive policy on mandatory use of certain percentage of recycled materials in lieu of virgin material ii) Expeditious implementation of EPR framework iii) Rebate in tax/GST on recycled products to increase its competitiveness
2.	Wet waste	Effective source segregation of waste: (i) Unbundling of sanitary landfills from SWM functions (ii) Relaxation of environmental clearance for waste processing plants (iii) Incentivizing of biogas plants through SATAT
3.	Construction and demolition waste	Implementing a comprehensive strategy and action plan covering the life cycle of construction projects, including dismantling phase: (i) Reduction in virgin construction of raw material usage in different building projects (ii) Extending tax rebates on recycled C&D products
4.	Wastewater	Creation of adequate sewage treatment capacity to meet the requirement of sewage generation in a time-bound manner. Targets for recycling and reuse of treated wastewater in the short, mid, and long term, at 25% by 2026, 35% by 2036, and 50% by 2050 respectively: (i) Preparation of new standards for designated reuse (ii) Framing of wastewater reuse policy by states/union territories (iii) Creation of institutional mechanism to promote circular economy in wastewater (iv) Mandatory use of recycled water in industries, especially in thermal power plants
5.	Municipal sludge	Treated sludge can be utilized as a resource: (i) Introducing national policy on sludge reuse/recycle (ii) Introducing comprehensive standards on recycle and reuse of processed sludge (iii) Incentivizing tagging of compost with chemical fertilizers and biogas with SATAT

4.3 Industrial symbiosis and waste recovery in India

Industrial symbiosis has become a popular term in recent years to describe industrial activities where a waste or by-product of one actor becomes a resource for another actor, that is, between industrial facilities or companies in which the waste or by-products of one become raw materials for another. Due to these, industrial symbioses become an increasingly salient issue in the field of industrial ecology and green growth. Recovery, reuse, and recycling of industrial residuals, often dismissed as wastes, are common in India mainly due to lower associated costs. Direct interfirm reuse is the cornerstone of the phenomenon, where firms cooperate in the exchange of material and energy resources, especially in industrial clusters and this is new to India.

The Naroda by-product exchange network provides the foremost example of eco-industrial development in India [52]. The Naroda industrial estate houses approximately 700 companies in a 30 km^2 region in Ahmedabad, Gujarat. Suggestions on potentially beneficial industrial symbiosis initiatives [53] include:

i. converting spent acid with high concentrations of H_2SO_4 to commercial-grade $FeSO_4$;
ii. selling sun-dried chemical gypsum to cement manufacturers, replacing the need for disposal;
iii. reducing the hazardous content of iron sludge produced by dye manufacturing industries, so that it could be used by brick manufacturers, in addition to reducing the amount of iron sludge being produced; and
iv. converting approximately 100 tons per month of industrial food waste to biogas.

More than a decade later, however, activities other than a common effluent treatment plant are catching up. A planned pilot project would create a "waste exchange bank" to facilitate the future exchange of residuals across companies.

Another study by the National Institute of Industrial Engineering, Mumbai, at the Taloja Industrial Estate, Raigad District, Maharashtra, revealed two existing by-product exchanges:

i. scrubbed tail gas from a petrochemical industry sold to another petrochemical industry for the manufacture of maleic anhydride and
ii. brewery wastes sent to a neighboring chicken farm for use as poultry feed [54]. The study identified six other potential avenues for symbiotic exchanges involving chemical solvents used by industries in this estate, but no reports of progress beyond this stage have been found.

Another study applied to material flow analysis in an economically diverse industrial area in South India characterizes the recovery, reuse, and recycling of industrial residuals. It quantifies the generation of waste materials from 42 companies as well as the materials that are directly traded across facilities and those that are recycled or disposed. This work encompasses a business cluster in Mysore, Karnataka, and is the first in India to thoroughly quantify material flows to identify existing symbiotic connections in an industrial area. Examined industries in this industrial area generate 897,210 MT of waste residuals annually and recovered 99.5% of these, and 81% was reused by the companies that generated them. Geographic data show that operations within 20 km of the industrial area receive over 90% of residuals exiting facility gates. Two-thirds of this amount goes directly to other economic actors for reuse. There were high rates of material and energy recovery in the Nanjangud Industrial Area (NIA). Companies in the uncovered network reuse their own by-products, share by-products with other local companies, and engage with local informal recycling markets. Symbiotic exchange among local companies was the second most important reuse strategy, while informal recycling accounted for most of the remainder of materials, and less than 1% was actually disposed. The sugar refinery produces 87% of the total residuals recovered; other configurations of industrial companies may not have this sort of lead player. From a thermodynamic standpoint, the NIA produces an impressively small amount of disposed waste. Further analysis could calculate the environmental and economic life cycle costs and benefits of the system. The high levels of symbiosis and recycling found reflect that these resources are valuable and that actors in both the formal and informal sectors capture and recycle these materials. As an early study of this kind in India, these results can build a foundation for a broader body of knowledge covering symbiosis, reuse, and recycling practices in India. Such studies can serve as a building block for companies to improve collaborative waste management practices and highlight opportunities for governments to guide and support these efforts.

4.4 Course on circular economy

Echoing the honorable prime minister of India's vision of "Circular Economy Mission," and motivated by the recent reforms carried out by the government, the AICTE – the nation's technical education controlling body – has realized the importance and need for inclusion of CE as a credit point course (4 credits) and introduce CE as an open elective course among young engineers and technologists (the syllabus of the course is provided below).

Syllabus of the circular economy course
CIRCULAR ECONOMY: A CREDIT POINT COURSE

Course code: CE
Course title: Circular economy
Number of credits: 4 (L: 2; T: 1; P: 2)
Course category: Open elective
Prerequisites: None

Course objective:
1. To develop graduates who have the necessary theoretical, practical, and research knowledge, skill, and aptitude in circularity and can get job opportunities by the industry in various sectors both public and private at national and international levels.
2. To contrive skilled manpower and entrepreneurship in the field of circular economy.
3. To enhance interaction of students with the senior/experienced manpower who have real time knowledge/experience in the technology development, research, innovation, entrepreneurship deployment, and circular business models.
4. To acquaint students about the needs of businesses related to circularity and to create zeal among students to pursue research and development (R&D) and entrepreneurship in this domain.
5. Create entrepreneurs who would promote knowledge in core competencies of environmental education and work on "innovation to industry" approach through university–industry partnerships.

Course content: (30 h)
Module I: Introduction to circular economy (4 h)
Linear economy and its emergence, economic and ecological disadvantages of linear economy, replacing linear economy by circular economy, development of concept of circular economy, a differential – linear versus circular economy

Module II: Characteristics of circular economy (4 h)
Material recovery, waste reduction, reducing negative externalities, explaining butterfly diagram, concept of loops

Module III: Circular design, innovation, and assessment (8 h)
Zero waste: Waste management in context of circular economy, circular design, research and innovation, LCA, circular business models

Module IV: Case studies (9 h)
Business models, solid waste management/wastewater, plastics: a case study, EPR: polluters pay principle, industrial symbiosis/eco-parks

Module V: Legal and policy framework (5 h)
Role of governments and networks, sharing best practices, universal circular economy policy goals, India and CE strategy, ESG

Learning resources:
Textbooks:
1. *The Circular Economy: A User's Guide*, Walter R. Stahel; Routledge, 1st Edition (24 June 2019)
2. *Circular Economy: (Re) Emerging Movement*, Shalini Goyal Bhalla; Invincible Publisher
3. *The Circular Economy Handbook: Realizing the Circular Advantage*, Peter Lacy, Jessica Long, Wesley Spindler; Palgrave Macmillan, UK
4. *Waste to Wealth: The Circular Economy Advantage*, Peter Lacy, Jakob Rutqvist; Palgrave Macmillan

References:
1. *Towards Zero Waste: Circular Economy Boost, Waste to Resources*, María-Laura Franco-García, Jorge Carlos Carpio-Aguilar, Hans Bressers; Springer International Publishing 2019
2. *Strategic Management and the Circular Economy*, Marcello Tonelli, Nicolo Cristoni; Routledge 2018
3. *Circular Economy: Global Perspective*, Sadhan Kumar Ghosh; Springer, 2020

4. *The Circular Economy: A User's Guide*, Stahel, Walter R.; Routledge 2019
5. *An Introduction to Circular Economy*, Lerwen Liu, Seeram Ramakrishna; Springer, Singapore 2021

Online resources:
1. https://www.coursera.org/learn/circular-economy
2. https://www.edx.org/course/circular-economy-an-introduction
3. https://www.coursera.org/learn/sustainable-digital-innovation
4. https://online-learning.harvard.edu/course/introduction-circular-economy?delta=0
5. https://www.oecd.org/cfe/regionaldevelopment/Ekins-2019-Circular-Economy-WhatWhy-How-Where.pdf
6. https://ic-ce.com/product/principles-of-circular-economy/
7. https://ic-ce.com/product/circular-business-management/
8. https://ic-ce.com/product/bootcamp/
9. http://ic-ce.com/journal-on-circular-economy/

Course outcomes:
At the end of the program students will be able to:
CO1 Apply the concept of circular economy to environmental engineering problems
CO2 Understand the concept of circularity and conduct relevant research
CO3 Use the principles of circularity for application to sustainable development
CO4 Apply complexity aspects of circular economy for creating circular business models

---

This course on CE is designed to improve knowledge studies in higher education by stimulating interactions between different approaches. This course would include
1. Knowledge on the subject
2. Research and development
3. Emerging innovation systems
4. Utility of knowledge production
5. Entrepreneurship development

With its comprehensive overview and multidisciplinary perspective, this CE course would provide/update the scholars with the theoretical and R&D-based information to make more informed decisions to protect our environment.

4.5 Conclusion

The need for protection and conservation of environment and sustainable use of natural resources is reflected in the constitutional framework of India and also in the international commitments of India. All industries need upgradation and significant structural changes to meet the expected market demand by improving resource effectiveness and reducing impact on environment by practicing concepts of CE and sustainable development. CE concept is useful to optimize the use of resources, reduce the exploitation of natural resources, and minimize the environmental pollution.

India's growing middle-income class, urbanization, and industrialization are major drivers of resource consumption. Although most resources are extracted domestically, India remains highly dependent on critical raw materials, which are important for its long-term development. India has made important progress in moving toward the sustainable development quadrant by releasing overarching and sectoral strategies on RE and CE. However, the planned strategies are yet to be implemented fully and this will require concerted efforts from various stakeholders in order to be successful and make India as resource efficient.

References

[1]   Ellen MacArthur Foundation and Material Economics, 2019. (Accessed on Aug 6, 2022)
[2]   Bennett, J.W., Pearce, D.W. and Turner, R.K., Economics of Natural Resources and the Environment. Johns Hopkins University Press, Baltimore, MD, USA. 1990; Am. J. Agric. Econ 1991, vol. 378, 227–228.
[3]   Census of India. Rural and Urban Distribution of Population. Government of India, New Delhi. 2011. (Accessed on May 2022)
[4]   Paraschiv, D., McDowall, R. and Thurlow, C., Energy efficient single detached dwellings beyond Ontario building code. Proceedings of 8th International Conference on Energy and Environment: Energy Saved Today Is Asset for Future, CIEM 2017. (Accessed on June 20, 2022, https://doi.org/10.1109/CIEM.2017.8120800)
[5]   Julkovski, D.J., Moslinger, E.A., Pereira, A. and Sehnem, S., Sustentabilidade na gestão e produção de ligas de chumbo a partir da reciclagem de baterias inservíveis. South American Development Society Journal, 2020, (Accessed on April 3, 2022 https://doi.org/10.24325/issn.2446-5763.v6i16p142-163)
[6]   Krill, M., Thurston, D.L., Guidat, T., Seidel, J., Kohl, H., Seliger, G., Zhao, S., Zhu, Q., Saavedra, Y.M.B., Barquet, A.P.B., Rozenfeld, H., Forcellini, F.A., Ometto, A.R., Matsumoto, M., Chinen, K., Endo, H., Yyyyy., W., Mendis, G.P., Peng, S. and Müller, D.B., Engine remanufacturing and energy savings Sahil. Journal of Cleaner Production, 2017, (Accessed on July 24, 2022 https://doi.org/10.1021/es300648w)
[7]   Zhang, A., Li, M., Lv, P., Zhu, X., Zhao, L. and Zhang, X., (2016) Disposal and reuse of drilling solid waste from a massive gas field. Procedia Environmental Sciences, 2016, 31, 577–581.
[8]   Foundation, E.M., Growth Within: A Circular Economy Vision for a Competitive Europe. Ellen MacArthur Foundation, 2015.
[9]   Urbinati, A., Chiaroni, D. and Chiesa, V., Towards a new taxonomy of circular economy business models. Journal of Cleaner Production, 2017, 168, 487–498.
[10]   NITI Aayog. 2021. (Accessed on May 10, 2022, https://pib.gov.in/PressReleasePage.aspx?PRID=1705772)
[11]   NITI Aayog. Strategy on resource efficiency. 2017. (Accessed on July 20, 2022, https://www.niti.gov.in/writereaddata/files/document_publication/Strategy%20Paper%20on%20Resource%20Efficiency.pdf)
[12]   NITI Aayog. Strategy on resource efficiency in steel sector. 2019a. (Accessed on May 24, 2022).
[13]   NITI Aayog. Strategy on resource efficiency in aluminium sector. 2019b. (Accessed on July 6, 2022).
[14]   NITI Aayog. Strategy on resource efficiency in construction and demolition sector. 2019c. (Accessed on August 5, 2022).
[15]   NITI Aayog. Strategy on resource efficiency in electrical and electronic equipment sector. 2019d. (Accessed on July 8, 2022).

[16] NITI Aayog. Resource efficiency and circular economy – current status and way forward. 2019e. (Accessed on August 6, 2022).

[17] National Resource Efficiency Policy (Draft) – Charting a Resource Efficient Future for Sustainable Development, Ministry of Environment, Forest and Climate Change Government of India, 2019.

[18] Kiiskila, K., Junnilainen, L. and Harkonen, M., Helsinki region congestion charging. Assessment of social impacts. In liikenne- ja viestintaministerion julkaisuja, publications of the ministry of transport and communications, 2011.

[19] TERI. Reference Report for National Resource Efficiency Policy for India. Prepared for MoEF & CC. Energy and Resources Institute, New Delhi, India. 2019. (Accessed on April 24, 2022 https://www.teriin.org/sites/default/fles/2019-04/National-Policy-Report.pdf)

[20] Biswas, A., et al. Waste-Wise Cities: Best Practices in Municipal Solid Waste Management. Centre for Science and Environment and NITI Aayog, New Delhi. 2021.

[21] MoHUA. Ministry of Housing and Urban Affairs, Government of India, circular economy in municipal solid and liquid waste. 2021. (Accessed on September 2, 2022, https://mohua.gov.in/pdf/627b8318adf18Circular-Economy-in-waste-management-FINAL.pdf)

[22] Bakshi, P., Pappu, A. and Gupta, M.J., A review on calcium-rich industrial wastes: A sustainable source of raw materials in India for civil infrastructure – Opportunities and challenges to bond circular economy. Journal of Material Cycles and Waste Management, 2021, 24(2), 49–62.

[23] Eric Williams. Achieving Climate Goals by Closing the Loop in a Circular Carbon Economy. King Abdullah Petroleum Studies and Research Center, King Abdullah Petroleum Studies and Research Centre (KAPSARC). 2019.

[24] IEA. 2020 – International energy agency (Accessed on August 23, 2022)

[25] https://ewastemonitor.info/wp-content/uploads/2020/11/GEM_2020_def_july1_low.pdf (Accessed on Sep 15, 2022).

[26] Electricals and Electronics manufacturing in India, ASSOCHAM, NEC technologies, 2018.

[27] CPCB. 2022. (Accessed on Sep 15, 2022, https://cpcb.nic.in/uploads/Projects/E-Waste/List_of_E-waste_Recycler.pdf).

[28] www.ewasteindia.com (Accessed on Sep 4, 2022)

[29] https://www.cerebracomputers.com/ (Accessed on Sep 14, 2022)

[30] https://www.attero.in (Accessed on Sep10, 2022)

[31] Haider, S., Adil, M.H. and Mishra, P.P., Corporate environmental responsibility, motivational factors, and effectiveness: A case of Indian iron and steel industry. Journal of Public Affairs, 2020, May 2020, 20(2), e2032.

[32] CPCB. 2007. (Accessed on Aug 1, 2022, https://cpcb.nic.in/openpdffile.php?id=UmVwb3J0RmlsZXMvTmV3U3RRlbV84OV8yNy5wZGY=)

[33] Ranitec. (Accessed on Aug 6, 2022, http://www.ranitec.com/pretreatmentsystem.php#:~:text=Tanneries%20doing%20Chrome%20Tanning%20Process,by%20adding%20magnesium%20oxide%20solution)

[34] CPCB. 2019. (Accessed on May 5, 2022, https://cpcb.nic.in/openpdffile.php?id=TmV3c0ZpbGVzLzcyXzE1NTQ0NTY2NDhfbWVkaWFwaG90bzIyMTYwLnBkZg==)

[35] SBM. 2020. Swachh Bharat Mission urban. http://swachhbharaturban.gov.in/ (Accessed on April 22, 2020).

[36] PIB. 2021: (Accessed on August 14, 2022) https://pib.gov.in/Pressreleaseshare.aspx?PRID=1779680; Joshi, R., and S. Ahmed. 2016. Status and challenges of municipal solid waste management in India: a review. Cogent Environmental Science 2:1139434.

[37] Planning Commission Report. Reports of the task force on waste to energy (Vol-I). (in the context of Integrated MSW management, Planning Commission, Government of India. 2014. (Accessed on August 20, 2022 http://planningcommission.nic.in/reports/genrep/rep_wte1205.pdf)

[38] Sri Shalini, S., Joseph, K., Yan, B., Karthikeyan, O.P., Palanivelu, K. and Ramachandran,, Solid waste management practices in India and China – Sustainability issues and Opportunities. Pardeep Singh, Yulia Milshina, Kangming Tian, Anwesha Borthakur, Pramit Verma and Ajay Kumar (ed), Waste Management Policies and Practices in BRICS Nations. CRC Press of Taylor & Francis, UK, 1st edition, 2021, 73.

[39] DRDPR – Rural Development and Panchayat Raj Department, Government of Tamil Nadu, 2020. Liquid waste management systems in rural Tamil Nadu, SBM. (Accessed on August 5, 2022, https://swachhbharatmission.gov.in/SBMCMS/writereaddata/Portal/SLWM/TamilNadu_GWM_PPT.pdf)

[40] Sujal, and Gaon, S., Resource material for field trainers developed by UNICEF in consultation with the Department of Drinking water and Sanitation (DDWS), Ministry of Jal Shakti, Government of India, 2019. (Accessed on September 17, 2022, https://swachhbharatmission.gov.in/SBMCMS/writ ereaddata/Portal/Images/pdf/Sujal%20and%20Swachh%20Gaon_5-day%20Manual%20(6% 20Sept).pdf

[41] BanasDairy. Banas Biogas plant, SBM. 2020. (Accessed on August 4, 2022, https://swachhbharatmission.gov.in/SBMCMS/writereaddata/Portal/SLWM/BanasDairy.pdf)

[42] NDDB- National Dairy development board, 2020, Efficient manure value chain, SBM. (Accessed on September 2, 2022, https://swachhbharatmission.gov.in/SBMCMS/writereaddata/Portal/SLWM/Ef ficientManureValueChain.pdf)

[43] Datta, P., Mohi, G.K. and Chander, J., Biomedical waste management in India: Critical appraisal. Journal of Laboratory Physicians, 2018, 10(1), 6–14.

[44] CPCB. Central Pollution Control Board. Status of sewage in India. Govt. of India, 2019. (Accessed on August 20, 2022, http://www.indiaenvironmentportal.org.in/files/file/revised-standards-STPs-NGT-Order.pdf)

[45] FSSM. National Policy on Faecal Sludge and Septage Management (FSSM), Government of India Ministry of Urban Development. 2017. (Accessed on September 1, 2022, http://amrut.gov.in/upload/ newsrelease/5a5dc55188eb0FSSM_Policy_Report_23Feb.pdf)

[46] FSSM. Draft policy on Faecal Sludge & Septage Management (FSSM), Government of Rajasthan. 2018. (Accessed on September 15, 2022, https://urban.rajasthan.gov.in/content/dam/raj/udh/organi zations/ruidp/MISC/FSSM%20Policy.pdf)

[47] FSSM. Faecal sludge and septage management in urban areas service & business model, NITI Aayog, 2021. (Accessed on September 4, 2022, https://www.niti.gov.in/sites/default/files/2021-08/ NITI-NFSSM-Alliance-Report-for-digital.pdf).

[48] PWMR. Plastic waste management amendment rules, Ministry of Environment, Forest and Climate Change Notification, New Delhi. 2022. (Accessed on September 6, 2022, https://cpcb.nic.in/uploads/ plasticwaste/2-amendment-pwmrules-2022.pdf)

[49] Franzen, H. Almost all plastic in the ocean comes from just 10 rivers, Deutsche Welle. 2017. (Accessed on July 16, 2022, https://www.dw.com/en/almost-all-plastic-in-the-ocean-comes-from-just-10-rivers/a-41581484).

[50] PWM. Plastic waste management. 2021. (Accessed on May 12, 2022, https://swachhbharatmission. gov.in/SBMCMS/writereaddata/Portal/Images/pdf/PWM_Manual_English_InnerPages.pdf)

[51] Saraswat, N., Transforming Industrial Estates in India. eNREE: A Quarterly Electronic Newsletter on Renewable Energy and environment. New Delhi. The Energy and Resources Institute, India. 2008, 2–5.

[52] Lowe, E.A., Eco-industrial Park Handbook for Asian Developing Countries. In: A Report to Asian Development Bank. Environment Department, Indigo Development, Oakland, USA. 2001.

[53] Unnikrishnan, S., Naik, N. and Deshmukh, G., Eco-industrial estate management: A case study. Resources, Energy, and Development, 2004, 1, 75–90.

[54] Bain, A., Shenoy, M., Ashton, W. and Chertow, M., Industrial symbiosis and waste recovery in an Indian industrial area, Resources. Conservation and Recycling, 2010, 54, 1278–1287.

Aline S. Tavares, Suzana Borschiver, and Claudio J. A. Mota

5 The circular economy in Brazil: cases and policies

Abstract: The Circular Economy is a holistic concept that is widespread across the world. It covers recycling, reuse, or remanufacturing, including circular design and circular business models. It is an opportunity to address negative environmental, economic, and social externalities with viable, sustainable solutions. In Brazil, the concept of circularity has been growing in various spheres of government, being included as an essential item on government agendas. In addition to bill number one, which establishes the National Circular Economy Policy and seeks the adoption of circularity in the Brazilian economy, the Fuels of the Future Program and the National Solid Waste Policy (PNRS), which established Reverse Logistics in the country, can be cited as practical actions to encourage circular processes, use of renewable energy, financing research in the area and sustainable public purchases. In this sense, this chapter aims to provide an overview of Brazilian public actions and policies related to the concepts and principles of the Circular Economy and their impact on the country. In addition to the measures mentioned above, other policies are also discussed, for example, in the energy sector, such as the "Incentive Program for Alternative Electric Energy Sources" (Proinfa) and RenovaBio, and in the waste management sector, such as the Circular Economy Route (REC) and the ERA-MIN Program. To this end, the country relies on the joint efforts of stakeholders such as the National Industry Confederation (CNI), the Ministry of Mines and Energy (MME), the Ministry of Economy (ME), the Ministry of the Environment (MMA), among others, whose participation is crucial.

5.1 Introduction

The world is facing scenarios caused by the human intervention in the nature, such as pollution, global warming, and loss of biodiversity, among others. Such events point to the urgent need to mitigate these negative externalities that characterize the traditional economic model called "take–make–dispose," which is reaching its limit.

In this sense, an alternative path with great potential for tackling this problem is the circular economy. This model seeks to associate growth and economic develop-

Aline S. Tavares, Suzana Borschiver, Universidade Federal do Rio de Janeiro, Escola de Química, Av. Athos da Silveira Ramos 149, CT bl E, Cidade Universitária, Rio de Janeiro 21941-909, Brazil
Claudio J. A. Mota, Universidade Federal do Rio de Janeiro, Instituto de Química, Av. Athos da Silveira Ramos 149, CT bl A, Cidade Universitária, Rio de Janeiro, 21941-909, Brazil; INCT Energia & Ambiente, UFRJ, Rio de Janeiro 21941–909, Brazil

https://doi.org/10.1515/9783110767179-005

ment with the preservation and improvement of the natural resources, optimizing the production process and the intelligent management of the renewable resources and finite feedstocks [1].

The circular economy proposes to create "an economic system that replaces the concept of end-of-life with reduction, reuse, recycling and recovery throughout the production, distribution, and consumption steps" [2]. Concomitantly, the circular model is based on the regeneration of the biological resources through biodegradation or composting, as well as the use of renewable energies (Figure 5.1).

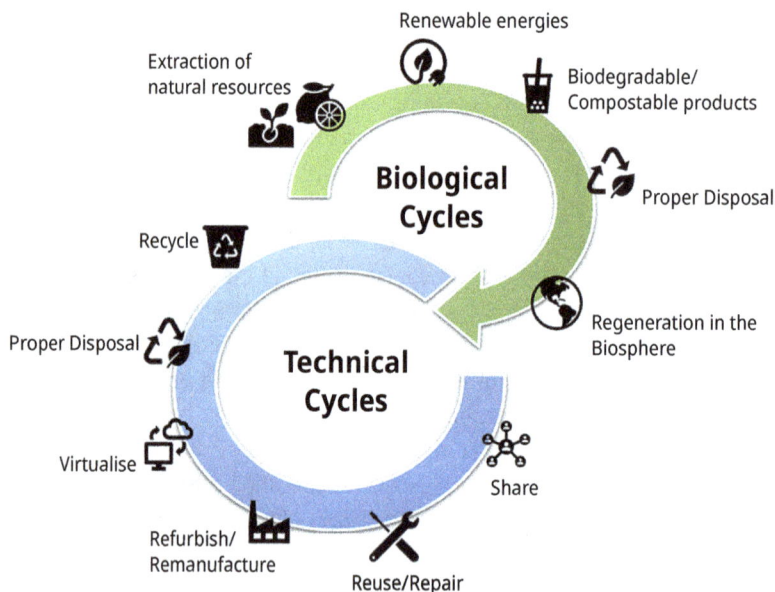

Figure 5.1: Biological and technical cycles of the circular economy. Source [3].

The circular economy proposes strategic planning from design, project, and manufacturing, for the transport, use, reinsertion in the production process and awareness of the relationship between human beings, raw materials, and wastes. However, only 9.1% of the world's economy is based on a circular model [4]. Thus, to gain more importance, it is necessary to create the enabling conditions for this transition, with the application, for example, of environmental education and adequate funding, as well as new disposal practices, public policies, infrastructure focused on circularity and innovative technologies [1].

To keep the average increase in the temperature of the planet within 1.5–2 °C until the end of the century, it is necessary to replace fossil energy sources by renewable ones, as well as significantly reduce the carbon emissions [5]. According to the International Energy Agency (IEA), it is estimated that the transition to a low-carbon economy will require around US$ 1 trillion in annual investments [6]. Thus, companies around the world are facing increasing pressure from stakeholders to expand initiatives toward more sustainable practices [7].

Thus, companies have sought in the proposal of the Environmental, Social and Governance (ESG) concepts, strongly related to the circular economy, as a way to leverage their investments. The effects of the COVID-19 pandemic and the growing demand from society for greater social justice, diversity, and inclusion has propelled companies to move up the ESG maturity scale at a faster rate [6, 8].

As a result, the circular economy brings new paradigms upon transforming how companies, government, and society produce, manage, and consume products and services, reconciling production and consumption with the ability to sustain natural resources and reduce wastes. In addition, inserting the circular economy in the global agenda is an opportunity to create a variety of new businesses and innovative technologies, rather than being just a sustainability issue [4].

5.1.1 Circular economy in the world

The concept of circular economy has been expanding among companies of different sectors. The Renault Group, for example, from the automotive sector, created a model for recycling end-of-life vehicles, called CAR REcycling 95% (ICARRE 95), supported by the EU LIFE Program [9]. The initiative seeks to reuse raw materials, such as steel, copper, and plastics, as well as metallic and textile wastes, performing the reverse logistics of these materials to manufacture new parts and enabling the practice of circular design. By keeping them as much as possible in the local automotive industry chain, the use of these components can still impact some sectors of the economy, such as petrochemicals and metallurgy, among others [10].

Circular design has played an important role in making the circular economy viable, by offering remodeled products after use. Another interesting case concerns Niaga, a Dutch joint venture of DSM, which has developed an adhesive that makes a material more easily separable, with the American company Mohawk, specialized in the manufacture of construction materials. Niaga sells fully recyclable carpets, where materials can be recovered after use and be transformed into new carpets or other materials [10].

With regard to international policies, the European Union (EU), China, the United States, and Canada are at the forefront of discussions for the adoption of a circular economy model, combining improvements to the ecosystem of extraction and use of raw materials to the reuse of wastes into profitable products [11].

The great diversity of wastes and material management across the EU, such as different recycling rates between countries, is one of the challenges faced by the bloc. On the basis of previous initiatives on waste management and resource efficiency, the policies were consolidated in the Circular Economy Action Plan in 2015 [12], which involves an economic package aimed at implementing the circular economy within the EU. For this purpose, more than 650 million euros were granted to projects that focused on promoting this model [13]. The main actions adopted were aimed at reducing the disposal of wastes in landfills and encouraging the reuse and recycling of wastes through economic incentives. Five priority areas were chosen: plastic, food wastes, essential raw materials, construction and demolition, and biomass and bio-based products [14].

In 2019, more challenging goals were set through the European Green Deal, to achieve neutral carbon emissions by 2050 in the bloc [15]. In addition to a climate-neutral economy, this agreement outlines strategies for financing environmentally friend projects aiming at becoming more resource-efficient and competitive [2, 15].

Regarding the situation in China, the country proposed, in 2005, the State Council of China to boost the circular economy, amid a national scenario of highly polluting industries. China also promulgated the outline of the National Program for Long- and Medium-Term Scientific and Technological Development (2006–2020) [12]. In 2009, it became the first country in the world to implement a specific law on circular economy, the Circular Economy Promotion Law, which aims at the improvement of resource use efficiency, environmental protection, and promotion of sustainable development, based on the 3Rs principle ("reduce, reuse, and recycle"), where the implementation includes fiscal instruments, financial support, and regulations that adopt the principle of shared product responsibility [14].

In 2015, a new law was implemented, being the country's first attempt to integrate the economic, social, and environmental spheres. It was recognized as the country's most progressive and rigorous act to implement more responsibility and accountability in local governments and law enforcement agencies, as well as setting higher standards for businesses from producer to recycler. Thus, more than 100 environmental regulations and policies were enacted to cover the entire supply chain [12]. From these initiatives, China is able to recover about 20% of the total mining wastes and reallocate them competitively in the market [11].

The Netherlands is another great example of a pioneering country in advancing toward circular economy. In 2020, the country instituted the Waste to Resource Program (VANG), with a focus on reducing waste generation, in addition to partnership with research institutes and industries [14]. The country also applied taxes on resources such as fuel, water, energy, and waste, in addition to reducing fees on maintenance and repair services, stimulating more efficient use and demand for long-lasting products [16].

5.2 Constructing circular economy in Brazil

5.2.1 Practices and potential opportunities

According to the National Industry Confederation (CNI), Brazil meets the conditions to be a leader in circular economy, because it has a diversified industry, wide highly qualified consumer, and scientific market, in addition to be abundant in natural resources and biodiversity. Nevertheless, public policies and a governance system will be the determining factors for the country's leverage on the subject [17].

In this sense, the Federation of Paraná State Industries (FIEP) works together with the state government to create a specific public policy for the circular economy. It has been reported that one of the plans is the launch of the "thematic chamber of circular economy," which aims at bringing together representatives from the industry, academia, and government to advance this agenda in the state [17].

Another example to be highlighted is the "circular economy network," launched by the Federation of Minas Gerais State Industries (FIEMG). The initiative allows companies to develop a business model for reusing and exchanging resources. Negotiations range from laboratory services and assignment of professional hours to the sale of waste and water [17]. FIEMG has also previously developed the "Circular Economy Program in Industrial Districts," with companies located in the Metropolitan Region of Belo Horizonte. Among the results of this program, it is worth mentioning 1.93 million ton of waste recycled per year, which resulted in the reduction of R$ 1.77 million in costs and 139.44 ton of carbon equivalent per year. Both are the results of the Minas Gerais Industrial Symbiosis Program, implemented in 2009 in the municipalities of Sete Lagoas and Uberaba [17].

A CNI[1] survey showed that 56.5% of the Brazilian industry has been working on process optimization to reduce wastes. In addition, the industries are advancing in practices that drive reverse logistics, such as the use of recycled or remanufactured materials [18]. Reverse logistics is a model that aims to plan, operate, and control the flow of post-sale and post-consumer goods, with a focus on returning to the business or production cycle to keep the components in the circular flow, for as many cycles as possible [19].

An industrial plant of Votorantim Cimentos, for example, collects and uses around 90,000 ton of açaí bagasse per year in its processes, reducing 100,000 ton of greenhouse gases, considering direct emissions [4]. Nespresso operates a recycling center since 2011 in the metropolitan region of São Paulo, where it developed a technology to separate coffee grounds and aluminum capsules. The sludge is sent to a composting center and transformed into organic fertilizer, whereas the aluminum returns to the industrial chain for recycling [4].

There is also a movement toward the business model called "product as a service," where companies maintain ownership of the product, whereas the customer pays for its functions or provided performance. The price per unit of product becomes the price per service, sharing responsibility for the product/service [20]. Virtualization, which deals with replacing physical assets by digital ones, and product sharing, both interrelated to the "product as a service," also appear in 16.5% and 15.9% of the cases, respectively, representing a total of 52.4% of the Brazilian industry in the servi-

---

1 The survey involved 1,261 industrial companies, chosen randomly, considering the national scenario.

tization of a consumer good and, by consequence, in its useful life extension and cost reduction [18].

5.2.2 Polices

5.2.2.1 Energy sector

The analysis of the circular model applied to the energy sector is of paramount importance, given the geopolitical and economic scenarios of energy transition and decarbonization. The Brazilian energy matrix is highly diversified, with renewable sources accounting for 44.1% of the share in 2021, characterized by the structural advances in wind and solar energy [21]. Therefore, this section will describe the policies that Brazil is being implementing toward the circular economy or related concepts within the energy sector.

– Distributed power generation
In 2002, the "Incentive Program for Alternative Electric Energy Sources" (Proinfa) was created and coordinated by the Ministry of Mines and Energy (MME). Focusing on distributed energy generation, the law was relevant for the Brazilian electric sector upon promoting the diversification of the energy matrix, the use of renewable energy, and the reduction of greenhouse gas emissions, contributing to the circular model [14].

This program also acted as a new stimulus for the expansion of biogas in the country, since it aimed to expand the generation of electricity through small hydroelectric plants, wind, and biomass sources. Once distributed generation was implemented, the surplus of electricity generated in rural regions could be inserted into the country's distribution networks. The great flexibility in the size and scale of the biodigesters also enhances the shared generation of electricity [22].

In 2012, two regulatory frameworks were published. The first was the "Electricity Distributed Generation Development Program" (ProGD), with the aim at stimulating energy generation by the consumers themselves, based on renewable sources and photovoltaics. The second deals with the Energy Compensation System in Brazil and establishes the concept of microgeneration, playing an important role in the operation of small residential power plants in the electrical distribution network [14, 23].

In addition to the programs mentioned before, subsidies were granted by the government. The first was the waiver of the ICMS (tax on circulation of goods) for electricity and distributed generation; the second was the exemption of IPI (tax on industrialized products) for equipment and components for harnessing solar and wind energy. The third incentive was the financing of distributed generation projects in public schools and hospitals for the development of renewable energies, including biogas [14]. Since 2011, for example, projects of biogas plants from agricultural wastes have received a total of BRL 566.8 million in investments [22].

– **Renovabio**

The National Biofuels Policy (RenovaBio) was instituted in 2017 and seeks to contribute to meeting the Brazilian goals under the Paris Agreement, promoting the expansion of production and use of biofuels in the national energy matrix. The law includes the use of instruments, such as targets for reducing greenhouse gas emissions in the fuel matrix, decarbonization credits, biofuels certification, compulsory additions of biofuels to fossil fuels and tax, financial, and credit incentives [24].

The targets for the reduction of the emissions were established in 2019 and are applicable up to 2029. Thus, the targets for decarbonization in the fuel sector encourage the increase of production and participation of biofuels in the transport energy matrix. The legislation also provides fines, civil, and/or criminal sanctions for cases of noncompliance with the individual annual emission reduction target [24].

– **Fuel of the Future Program**

The National Energy Policy Council (CNPE) instituted the "Fuel of the Future Program" in 2021, and created the Fuel of the Future Technical Committee (CT-CF), composed of 15 institutions: Ministry of Mines and Energy (MME); Civil Secretary of the Presidency of the Republic; Ministry of Economy (ME); Ministry of the Environment (MMA); Ministry of Infrastructure (MInfra); Ministry of Agriculture, Livestock and Supply (MAPA); Ministry of Foreign Affairs (MRE); Ministry of Science, Technology and Innovation (MCTI); Ministry of Regional Development (MDR); Brazilian Maritime Authority; National Agency of Petroleum, Natural Gas and Biofuels (ANP); National Civil Aviation Agency (ANAC); Energy Research Company (EPE); Brazilian Institute for the Environment and Renewable Resources (IBAMA); National Institute of Metrology, Quality and Technology (Inmetro) [25].

The program was created with the objective of expanding the use of sustainable and low-carbon fuels in the country, including the following specific objectives [25]:

- Integration of public policies: RenovaBio, National Program for the Production and Use of Biodiesel, Proconve, Route 2030, Brazilian Vehicle Labeling Program and CONPET;
- Proposing a complete life cycle analysis methodology (from the well to the wheel) in the different transport modals;
- Providing the consumer with adequate information in order to contribute to the conscious choice of vehicle and energy source;
- Proposing specification studies for high-octane gasoline and ethanol fuel cell;
- Introduction of sustainable aviation fuel (SAF) in the Brazilian energy matrix;
- Establishing a strategy for the use of sustainable fuels in national maritime transport;
- Use of carbon capture and storage technologies associated with the production of biofuels.

5.2.2.2 National Solid Waste Policy (PNRS) and reverse logistic

The National Solid Waste Policy (PNRS) is the first public policy in the country with a more specific focus on waste management [14]. In addition to the prevention and reduction of waste generation, the legislation also presents the practice of sustainable consumption habits, the increase in the recycling rate, and reuse of solid wastes as important factors for tackling its inappropriate disposal [24].

In this context, the same policy instituted the reverse logistics model, described as "a set of actions, procedures and means designed to enable the collection and return of solid wastes to the business sector, for reuse, in its cycle or in other production cycles, or other environmentally appropriate final destination." Cases of mandatory application of reverse logistics were then established for the following products [26]:

I   – pesticides, their residues and packages;
II   – batteries;
III – tires;
IV   – lubricating oils, their residues and packages;
V   – fluorescent, sodium and mercury vapor, and mixed light lamps;
VI   – electronic products and their components.

Today, reverse logistics is well structured for lubricating oils and batteries, since these sectors already have several recycling plants with technology at similar stages to those in Europe [24]. The forecast is to extend the program to plastic, metal and glass packages, in addition to other packages, considering the impact on public health and the environment [26].

The policy also includes the concepts of shared responsibility for the life cycle of the product within the entire chain, and the recognition of reusable or recyclable solid wastes as an asset of economic value, including manufacturers, importers, distributors, and traders. Actions from the public authorities are also provided to meet initiatives to prevent and reduce the generation of solid wastes in the production process [14].

However, the implementation of the law still faces challenges. The mandatory implementation of reverse logistics for all types of waste has also been discussed, having a transition period for the market and consumers, with the public sector being in charge of the regulation [26].

For the reverse logistics to be viable in Brazil, it is necessary to reduce bureaucracy in administrative processes to facilitate the transit of collected materials. Another limiting factor that has been pointed out deals with the tax cumulativeness of the reuse chain, which waiver is reported as a fundamental factor to make the model more feasible [17].

On the other hand, some advances can be pointed out. In 2019, the Ministry of the Environment launched the "Zero Garbage Program" and the "Plan to Combat Garbage in the Sea." The first has the objective of eliminating dumps and to support the municipalities in finding solutions for the final disposal of solid wastes, whereas the second has the objective of creating an overview of garbage in the sea in Brazil, forming reference values, the desired scenario, a governance model, guidelines, indicators, action plan, and agenda of the activities [24].

In 2020, the Ministry of the Environment carried out a public query on the term of commitment for the implementation of actions focused on the circular economy and reverse logistics of packaging in general, called the ReCircula Initiative. The term of commitment involves six large companies, mostly from the food sector (Ambev, Kaiser & HNK BR, Nestlé, Coca-Cola, Tetrapak, and Unilever) and establishes a target of 100% of plastic packaging made of recyclable or compostable materials by 2025 [24].

Among the Brazilian states, Paraná is considered one of the pioneers in initiatives to advance reverse logistics. Since 2012, the Paraná government has launched calls for the industrial sectors to submit proposals focused on this model. In addition, since 2021, the legislation obliges nine sectors – among them, pesticides, light bulbs, packaging, and tires – to present information on reverse logistics on the "accounting for waste" platform [17].

In April 2022, the National Plan for Solid Waste (PLANARES) was instituted, and it aims at establishing guidelines, strategies, programs, projects, and actions for the final disposal of wastes, in addition to control its implementation and operationalization [24]. The plan was drawn up through a technical cooperation agreement between the Ministry of the Environment and ABRELPE (Brazilian Association of Public Cleaning and Special Waste Companies) and has been validated throughout the national territory [27].

5.2.2.3 Circular economy route (REC)

In 2019, with the aim of promoting the management of sustainable alternatives and stimulating the regional economic development, the circular economy route (REC) was structured by the Ministry of Regional Development (MDR). By the end of 2020, two poles had been structured: the Cerrado Circular (Integrated Development Brasília Region – Federal District – RIDE/DF) and the Paraíba Circular. Both initiatives created management committees and project portfolios with two projects running [1].

In partnership with the Ministry of Science, Technology and Innovations (MCTI) and the Ministry of Communications (MCom), the MDR defined the following areas of activity for the REC: (i) strengthening the recycling production chain with reuse, regeneration, and transformation of solid wastes; (ii) development of renewable energy management (wind, biomass, photovoltaic, heliothermic, thermal) for urban development; (iii) development of technologies to optimize the use of water resources (water reuse, desalination, reverse osmosis) [1].

With regard to actions aimed at the recycling production chain in the national territory, five action lines were defined [1]:

- Mapping of economic enterprises linked to the recycling chain, as well as support for the development of new technologies;
- Structuration of five poles, one in each macroregion of the country;
- Diagnosis and management of solid wastes in selected regions;
- Construction of an online platform for marketing waste.

In the field of renewable energy and solid waste management, the National Secretariat for Mobility and Regional and Urban Development (SMDRU)-MDR develops, in partnership with the Federal Institute of Education, Science and Technology of Brasília (IFB), a pilot project for the implementation of biodigesters demonstrative units in four campuses of the institute. The project began in March 2021 and has the following goals [1]:

- Training of medium- and high-level technicians in solid waste management and renewable energies, already concluded;
- Reduction in the disposal of organic wastes in condominiums and IFB campuses;
- Creation of at least two startup companies to develop circular economy solutions;
- Conducting a seminar to disseminate the results.

As for the third segment, focusing on the management of water resources, renewable energy, and solid waste management, SMDRU-MDR is developing a pilot project in partnership with the Federal Rural University of the Semi-Arid (UFERSA). The objective is the development and implementation of technologies for the semiarid region and presents the following goals [1]:

- Building a digital platform for distance learning;
- Dissemination of technologies through electronic material (catalogue, booklets, videos, and tutorials);
- Creation of at least two startup companies to develop solutions with an emphasis on the circular economy;
- Diffusion of technologies inserted in the platform.

5.2.2.4 ERA-MIN

ERA-MIN is a consortium formed by 25 funding organizations aimed at research, development, and innovation in the world, which includes the Brazilian Finep (Financier of Studies and Projects). Between 2011 and 2015 the first edition took place, whereas the second edition (ERA-MIN 2) is in execution since 2016. The third edition (ERA-MIN 3) is projected to take place between 2020–2025 [28, 29].

The objective of the consortium is to financially support transnational R,D&I projects that are jointly developed by Brazilian companies and Scientific and Technological Institutions (ICTs), with a focus on the circular economy in mineral raw materials and their secondary sources (metallics, construction, and industrial minerals), involving the supply chain, production, consumption, reuse, and recycling [28].

Finep has also launched the Startup Program in 2017. The objective is to support technology-based startup companies that are developing solutions on various topics, including the circular economy. Specifically in this segment, the program seeks innovations with a focus on eliminating pollution and wastes, as well as using materials and resources in a regenerative and restorative way [30].

The Rio de Janeiro State Foundation for Research Financing (FAPERJ) also has programs specifically devoted for the creation of startup companies within the Rio de Janeiro State. Among the several startups that were created within this initiative is CarbonAir Energy, which intends to provide solutions on CO_2 capture and utilization [31].

5.2.2.5 Sustainable policies related to circular economy

– Biotechnology Development Policy (PDB)
The Biotechnology Development Policy (PDB) and the National Biotechnology Committee were established in 2007. Among the objectives of the PBD is the development of innovative biotechnological products and processes, the stimulation to improve the efficiency of the productive structure, the increase in the innovation capacity of Brazilian companies, the absorption of technologies, and the expansion of the exports [24].

Concerning the industrial area, the policy emphasizes the inclusion of industrial and special enzymes to enhance the production of renewable fuels through more efficient biotechnological processes. This, in turn, favors the use of renewable raw materials and minimizes the generation of effluents and use of water resources [24].

The priority areas in the environmental sector are the treatment of wastes and effluents, treatment of environmental contamination, environmental recovery techniques, and conservation of species, among others. The policy also aims at promoting the development of products and processes for biopolymers, such as biodegradable plastics, from renewable resources, such as sugar cane, corn, among others [24].

– National Policy on Climate Change (PNMC)
The National Plan on Climate Change was launched in 2008. The National Policy on Climate Change (PNMC) was instituted and regulated in 2010. The law established plans for mitigation and adaptation to climate changes [24].

One of the plans was dedicated to agriculture: the Sectorial Plan for Mitigation and Adaptation to Climate Change for the Consolidation of an Economy of Low Carbon Emissions in Agriculture (ABC Plan) [22]. The law has the national fund on climate change, whose objective is to finance projects and studies aimed at reducing greenhouse gas emissions. Linked to the Ministry of the Environment (MMA), the law also provides means of financial and economic instruments, such as tax, carbon market, among others [24].

The Nationally Determined Contribution (NDC) was a Brazilian commitment, in the scope of the Paris Agreement, in the direction of developing actions to mitigate climate changes. The NDC includes a target of 37% reduction in greenhouse gases emissions for the entire economy by 2025 and 43% by 2030, based on the 2005 emissions. The National Bank for the Economic and Social Development (BNDES) is another Brazilian player that supports projects of relevant environmental impact. The Climate Fund, for example, can provide up to 80% of the total project budget [22].

Regarding the Brazilian states, the Paraná government created the "Climate Seal" for companies that publish their inventories [17]. The Paraná Climate Seal is an award for companies that voluntarily carry out the reduction, measurement, and disclosure of the carbon footprint [32]. The last edition was held in December 2021 and included 70 companies and business conglomerates that accounted 177 industrial units, distributed in 66 municipalities in Paraná state. The city hall of Maringá was the pioneer public organization to receive the Seal [33].

5.3 Challenges and barriers to implementing the circular economy in Brazil

From what has been previously described, it can be inferred that Brazil has taken important steps toward making the circular economy viable, albeit with some limitations and challenges to be overcome. At the international level, the country is involved in the preparation of technical standards on circular economy through the CNI, which are being coordinated by the International Organization for Standardization (ISO) and should come into force by 2023 [17].

There are still many existing bottlenecks. The existence of data gaps to identify value chains, as the entire census system is not adapted to the theme. Circular initiatives need to gain scale to increase the competitiveness of the products [24]. Another major limitation often mentioned for the implementation of the circular economy is concerned with the efficient waste management. In 2020, the amount of uncollected wastes was almost 7 million ton. In addition, more than 30 million ton of collected waste was dumped in inappropriate places (dumps and controlled landfills) in the same year [34].

Although public policies focused on the circular economy are being implemented in the country, mainly in the energy sector, it is clear that there is still lack of public policies aimed at areas such as cleaner production, climate protection, incentives for circular design, and waste, among others.

Despite the advances achieved with the reverse logistics of tires, lubricating oils, and batteries, as mentioned in item 5.2.2.2, there are still limiting factors in other sectors, such as light bulbs, because its reverse logistics is more expensive compared to a new lamp, with no technological alternative available yet [24]. Other obstacles deal with the recyclables collection system and the precarious logistics network, in addition to double taxation of recyclables, acting as a legal barrier [11, 24].

The bases for the expansion of the concept in Brazil demand innovation in efficient business models that accelerate the formation of new markets [24]. It is necessary to build an adequate regulatory framework, with tools and metrics adapted to the reality of Brazilian organizations and with the collaboration of the various players of the society (Figure 5.2) [17].

Promote society's awareness for the best use of resources.

Create a governance system for the Circular Economy, stimulating collaboration between institutions and building a national strategy for better management of natural resources.

Strategic management of the flow of resources.

Investments in research, development and technological innovation.

Encourage circular value chains.

Strengthen the competitiveness and expansion of the national manufacturing industry.

Figure 5.2: Actions for the advancement of the circular economy in Brazil.
Source: Own elaboration based on Ref. [17].

A governance system that encourages greater articulation and integration between institutions and investments in R,D&I can make circular value chains viable, enhance the competitiveness of the national industry, and, consequently, the systemic view of the final positive impact of its practices by companies. One can also add incentives for new technologies and design, in addition to the condition for the development of circular initiatives for small- and medium-sized companies.

5.4 Conclusion

It can be understood that the circular economy, as a new holistic model of production and services, has great potential to boost the Brazilian economy, in line with the goals of sustainable development, such as the mitigation of climate changes and the widespread use of renewable energy. Its principles reinforce the need for disruptive changes in the production process, consumption patterns and new public policies.

The "modus operandi" of this model also brings a scenario that proves to be challenging and, at the same time, presents itself as a strong alternative to revert the negative externalities of the traditional linear economic model. It should also be noted that this model does not just deal with reduction, reuse, and recycle, the 3Rs. The circular design of products assumes an important position, since it enables the separation and return of components to the production chain.

Furthermore, it is a fact that the industrial sector plays a fundamental role in this evolutionary process by adding value to natural resources and wastes, transforming them into products for the consumer market. However, the support of public policies has proved to be an important driver, as is the case in the EU, China, and the Netherlands, for example. In Brazil, a recent movement can be seen with the integration between the market and government, allowing the definition of priority sectors and long-term goals, as in the case of Fuels of the Future and Circular Economy Routes (REC) programs.

The circular economy shows promise in the transition to a low or zero carbon economy. Brazil emerges as a competitive and important player in a global circular economy scenario, as the country is strong in the agrobusiness and has great potential for renewable energy.

References

[1]   Rota da Economia Circular. Brasília, DF.: MDR – Ministério do Desenvolvimento Regional. 2021. (Accessed on November 22, 2022, at https://www.gov.br/mdr/pt-br/assuntos/desenvolvimento-regional/rotas-de-integracao-nacional/rota-da-economia-circular).

[2]   Holzer, D., Rauter, R., Fleiß, E. and Stern, T., Mind the gap: Towards a systematic circular economy encouragement of small and medium-sized companies. Journal of Cleaner Production NL, 2021, 298, 126696.

[3]   Tavares, A., Bandarra, R. and Borschiver, S., What is a circular economy?. Circular Economy: A New Mindset on Sustainable Development., Ed. Amazon Kindle Publishing, Rio de Janeiro, RJ, BRA, 1st edition, 2021, 20–31.

[4]   CEBDS – Conselho Empresarial Brasileiro Para O Desenvolvimento Sustentável. Economia Circular. Quebrando Muros. CEBDS, Rio de Janeiro, RJ, BRA, 2019.

[5]   Mota, C.J.A., Chagas, J.A.O. and Marciniakb, A.A., Trends in carbon dioxide capture and conversion. Journal of the Brazilian Chemical Society BRA, 2022, 33(8), 801–814.

[6]   Braga, C., Como o ESG está mudando o Mercado Financeiro? Sessão 2: ESG: Um Olhar Do Mercado Financeiro Para a Sustentabilidade., Fundação Dom Cabral (FDC). BRA, 2022, 3–6.

[7]   Hadro, D., Fijałkowska, J., Daszynska-Zygadło, K., Zumente, I. and Mjakuškina, S., What do stakeholders in the construction industry look for in non-financial disclosure and what do they get?. Meditari Accountancy Research ENG, 2021, 30(3), 762–785.

[8]   Nassos, G.P.N. and Avlonas, N., Practical Sustainability Strategies. How to Gain a Competitive Advantage. John Wiley & Sons, Inc, NJ, USA, 2nd edition, 2020.

[9]   INDRA. A Competitive End-of-life Vehicles Treatment Branch, Recuperates up to 95% of Car Mass. European Union, NL, 2019. (Accessed on November 22, 2022, at https://circulareconomy.europa.eu/platform/en/good-practices/indra-competitive-end-life-vehicles-treatment-branch-recuperates-95-car-mass)

[10]   Tavares, A.S. and Borschiver, S., Elaboration of technological and business roadmap of the circular economy. Cad de Prospec BRA, 2021, 14(3), 810–823.

[11]   Caminhos para expandir a economia circular no Brasil: Dos pequenos aos grandes negócios. Belo Horizonte, MG.: FUNDEP, 2021. (Accessed on November 22, 2022, at https://www.fundep.ufmg.br/economia-circular-no-brasil/).

[12]   Zeng, X. and Li, J., Circular economy in China. Circular Economy Global Perspective. Ghosh, SK, Springer Nature Singapore Pte Ltd, Singapore, 1st edition, 2020, 123–129.

[13] Closing the loop – An EU action plan for the circular economy COM/2015/0614 final. DK.: European Environment Agency, 2015. (Accessed on November 22, 2022, at https://www.eea.europa.eu/policy-documents/com-2015-0614-final).
[14] Economia Circular. Caminho Estratégico Para a Indústria Brasileira. CNI – Confederação Nacional da Indústria, Brasília, DF, 2019. (Accessed on November 22, 2022, at https://www.portaldaindustria.com.br/publicacoes/2019/9/economia-circular-caminho-estrategico-para-industria-brasileira/#circular-economy-strategic-path-for-brazilian-industry)
[15] A European Green Deal. EU.: European Commission. 2019. (Accessed on November 22, 2022, at https://ec.europa.eu/info/strategy/priorities-2019-2024/european-green-deal_en).
[16] Research & innovation projects relevant to the circular economy strategy: Calls 2016 – 2018: Horizon 2020. EU.: European Commission, 2019. (Accessed on November 22, 2022, at https://ec.europa.eu/info/sites/default/files/h2020_projects_circular_economy_2016-2018.pdf).
[17] Práticas de economia circular contribuirão para o Brasil atingir meta de neutralidade de carbono até 2050. Brasília, DF.: Portal da Indústria, 2021. (Accessed on November 22, 2022, at https://noticias.portaldaindustria.com.br/noticias/economia/praticas-de-economia-circular-contribuirao-para-brasil-atingir-meta-de-neutralidade-de-carbono-ate-2050/#).
[18] 76,4% das indústrias desenvolvem alguma iniciativa de economia circular, mostra pesquisa da CNI. Brasília, DF.: Portal da Indústria, 2019. (Accessed on November 22, 2022, at https://noticias.portaldaindustria.com.br/noticias/sustentabilidade/764-das-industrias-desenvolvem-alguma-iniciativa-de-economia-circular-mostra-pesquisa-da-cni/).
[19] Leite, P.R., Logística Reversa: Meio Ambiente E Competitividade. SP, BRA, Prentice Hall, 2009.
[20] Witjes, S. and Lozano, R., Towards a more circular economy: Proposing a framework linking sustainable public procurement and sustainable business models. Resources, Conservation and Recycling NL, 2016, 112, 37–44.
[21] EPE lança Relatório Síntese do Balanço Energético Nacional 2022, com informações consolidadas do setor energético no Brasil no ano de 2021. Brasília, DF.: EPE – Empresa de Pesquisa Energética, 2021. (Accessed on November 22, 2022, at https://www.epe.gov.br/pt/imprensa/noticias/epe-lanca-relatorio-sintese-do-balanco-energetico-nacional-2022-com-informacoes-consolidadas-do-setor-energetico-no-brasil-no-ano-de-2021).
[22] Milanez, A.Y., da S, Maia GP. and Guimarães, D.D., Biogás: Evolução recente e potencial de uma nova fronteira de energia renovável para o Brasil. BNDES Set, 2021, 27(53), 177–216.
[23] Geração Distribuída. Brasília, DF.: ANEEL – Agência Nacional de Energia Elétrica. 2022. (Accessed on November 22, 2022, at https://www.gov.br/aneel/pt-br/assuntos/geracao-distribuida).
[24] Economia Circular. Os desafios do Brasil. Rio de Janeiro, RJ.: CEBRI – Centro Brasileiro de Relações Internacionais. 2020. (Accessed on November 22, 2022, at https://cebri.org/br/doc/24/economia-circular-os-desafios-do-brasil).
[25] Combustível do Futuro. Brasília, DF.: MME – Ministério de Minas e Energia. 2021. (Accessed on November 22, 2022, at https://www.gov.br/mme/pt-br/assuntos/secretarias/petroleo-gas-natural-e-biocombustiveis/combustivel-do-futuro).
[26] Azevedo, J.L., A economia circular aplicada no brasil: Uma análise a partir dos instrumentos legais existentes para a logística reversa. XI Congresso Nacional de Excelência em Gestão, 2015, 1–16.
[27] Plano Nacional de Resíduos Sólidos – Planares apresenta os caminhos para que a gestão de resíduos avance no país, com a recuperação de 50% dos resíduos em 20 anos. São Paulo, SP.: ABRE – Associação Brasileira de Embalagens, 2022. (Accessed on November 22, 2022, at https://www.abre.org.br/sustentabilidade/plano-nacional-de-residuos-solidos-planares-apresenta-os-caminhos-para-que-a-gestao-de-residuos-avance-no-pais-com-a-recuperacao-de-50-dos-residuos-em-20-anos/).
[28] ERA-MIN 2. Rio de Janeiro, RJ.: FINEP – Financiadora de Estudos e Projetos. (Accessed November on 22, 2022, at http://www.finep.gov.br/cooperacao-internacional-externo/era-min-2).

[29] Raw materials for the sustainable development and the circular economy. ERA-MIN EU – ERA-NET confund on raw materials. (Accessed November 22, 2022, at https://www.era-min.eu/).

[30] Programa de Investimento em Startups Inovadoras. Rio de Janeiro, RJ.: FINEP – Financiadora de Estudos e Projetos. 2020. (Accessed on November 22, 2022, at http://www.finep.gov.br/chamadas-publicas/chamadapublica/637).

[31] CarbonAir Energy. (Accessed on November 24, 2022, at https://carbonairenergy.com.br/)

[32] Clima Paraná, S., Ponta Grossa, P.R.: Agrocete. 2021. (Accessed on November 22, 2022, at https://agrocete.com.br/pt/grap-no-campo/selo-clima-parana).

[33] Sétima edição do Selo Clima Paraná teve recorde de indústrias premiadas. Curitiba, PR.: Agência FIEP, 2021. (Accessed on November 22, 2022, at https://agenciafiep.com.br/2021/12/08/setima-edicao-do-selo-clima-parana-teve-recorde-de-industrias-premiadas/).

[34] Panorama dos Resíduos Sólidos no Brasil. 2021. São Paulo, SP.: Abrelpe – Associação Brasileira de Empresas de Limpeza Pública E Resíduos Especiais. 2022. (Accessed on November 22, 2022, at https://abrelpe.org.br/panorama/).

Kibeak Lee

6 The era of the circular economy driven by new technology for carbon neutrality: changes in South Korea

In a linear economic system, consumption-oriented economic development will eventually undermine sustainability. With such development, South Korea, lacking natural resources, may not be able to usher in a sustainable future. The transition to a circular economy system may require a considerable amount of cost and effort, but it could be a new means of addressing the chronic problem of limited resources in this country.

The circular economy is a key element of a sustainable future city aimed at enhancing environmental quality, achieving economic growth and recovery of a healthy society. To embody this notion, comprehensive strategies that take into account the environment, economy, and society are needed, and related knowledge and experience must be accumulated. South Korea has acknowledged these requirements and is promoting transformations in the country, businesses, and its citizens using a variety of policy tools.

6.1 Definition of circular economy

The circular economy refers to a system where materials and energy are circulated to minimize raw materials, waste, and energy losses, and is a concept that contrasts with the existing linear economy model of take–make–dispose [1, 2] (Figure 6.1). According to the European Commission, a circular economy is defined to maintain the value of products, materials, and resources for as long as possible, while minimizing the generation of waste [3].

In the policies of the Korean government, the meaning of the circular economy has been considered to be the same as resource circulation. However, its meaning has been expanded to an economic system for realizing a sustainable society in addition to resource circulation since the declaration of carbon neutrality [1]. Moreover, Kirchherr et al. defined a circular economy as a system that aims to achieve sustainable development at the microlevel (products, companies, consumers), mesolevel (eco-industrial parks), and macrolevel (cities, regions, nations, and beyond) [4].

Kibeak Lee, Korea Research Institute of Chemical Technology (KRICT), 141 Gajeong-ro, Yuseong-gu, Daejeon 34114, South Korea

https://doi.org/10.1515/9783110767179-006

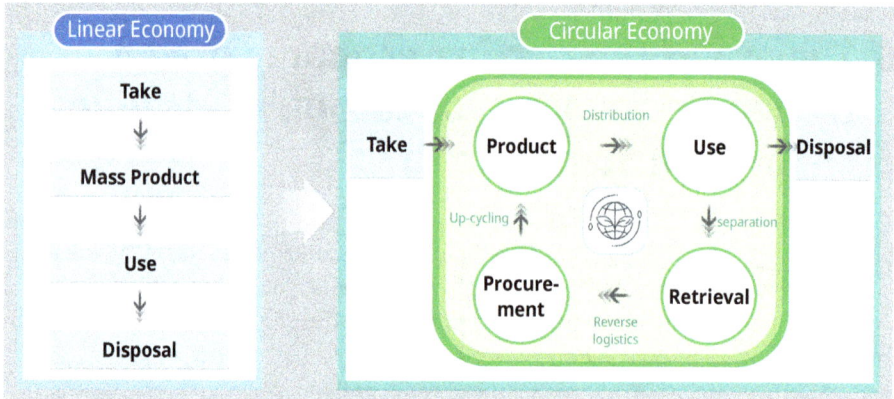

Figure 6.1: Linear and circular economy.

As such, the circular economy can be defined as a sustainable system that improves environmental quality, ensures economic growth, and achieves social equity, thereby benefiting both present and future generations. To realize a circular economy, new business models and responsible consumers are required.

6.2 Current status of South Korea

6.2.1 Climate and industrial environment

Over the last 100 years, the global average temperature has risen by 0.8–1.2 °C, while the average temperature in South Korea has risen by 1.8 °C [5]. The temperature in South Korea has increased by a significant 1.4 °C in the past 30 years, showing a much stronger global warming trend compared to other countries. This has led to intensified precipitation polarization and a high possibility of drought and flooding, resulting in many serious natural disasters [6].

As South Korea is located in the mid-latitude temperate climate zone, it has seasonal characteristics where each of the four seasons is distinctly different from one another, making it difficult to use renewable energy. As for wind power generation, 25% and 40% of wind resources are available for onshore and offshore wind power, respectively, and in the case of solar power, the solar radiation significantly varies by season, which makes it difficult to generate electricity from renewable energy evenly throughout the year [7]. The share of renewable energy-driven power generation in the United States is 17.4%, 39.9% in Germany, and 18.6% in Japan, while South Korea is at 4.9% [8] (Figure 6.2).

In terms of industrial structure, South Korea imports most of its energy resources and its industrial structure is centered on energy-intensive manufacturing industries such as steel, petrochemicals, and semiconductors. Furthermore, South Korea has an export-driven economic structure centered on manufacturing. As of 2019, the manufacturing industry accounted for 27.5% of GDP (cf. US 10.9%, EU 16.4%), and its dependence on exports remains high at 30.5% [9].

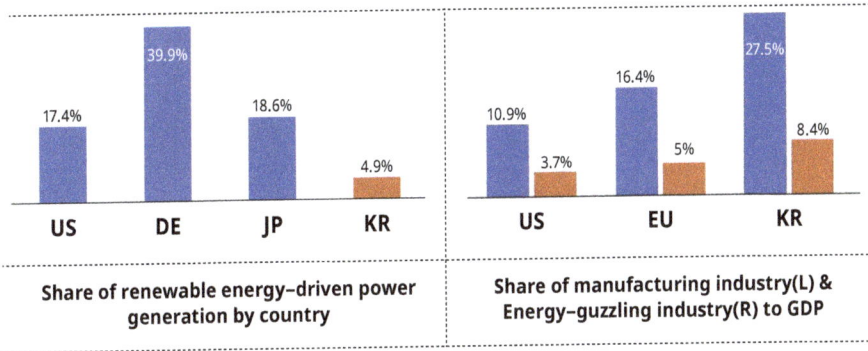

Figure 6.2: Share of renewable energy and manufacturing industry (2019).

The US and Europe have recently introduced export regulations and related policies such as carbon border taxes. Therefore, it is essential for South Korea, which ranks 7th in the world export of commodity trade [9], to respond to such global changes.

Furthermore, compared to advanced countries such as the US and Europe, the level of technology for industrial efficiency is only about 81.5% and the level of carbon-neutral technology including CCUS (carbon capture, utilization, and storage) technology is assessed to be about 80% [10], which requires continuous investments and interest in securing technologies (Figure 6.3).

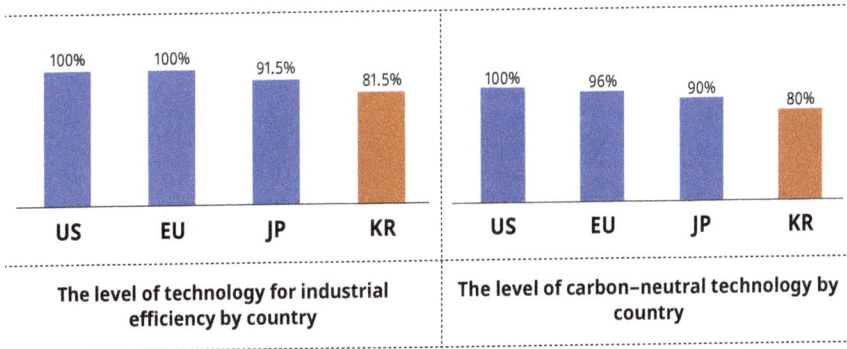

Figure 6.3: Current status of carbon-neutral technology (2020).

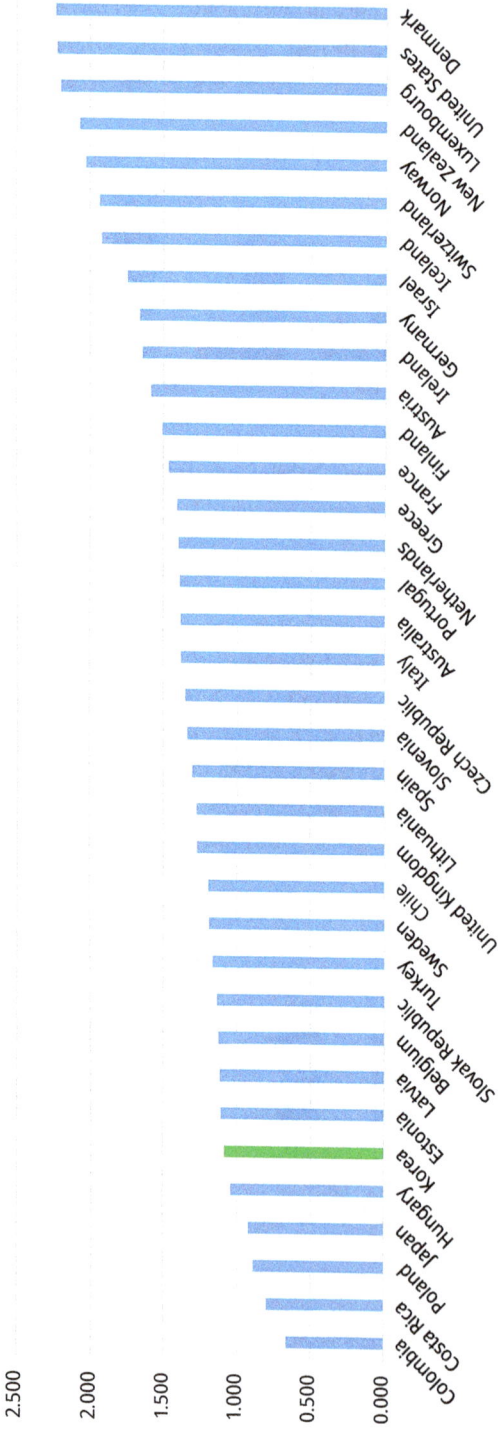

Figure 6.4: Average daily waste generated per capita by country (unit: kg/person/day, '18).

6.2.2 Waste management

Through its relevant policies, South Korea has shown an exemplary model for waste management, moving away from the growing amount of waste generation coupled with increasing consumption. As of 2018, the waste generation in South Korea stood at 1.09 kg of waste per capita per day, the sixth smallest amount among OECD countries (cf. US 2.22 kg, UK 1.27 kg, Japan 0.93 kg), and approximately 88% of the total waste generated was recycled in 2020 with the landfill rate steadily declining at 4.76% [9] (Figure 6.4).

Despite the difficult situation with problems such as resource depletion and environmental pollution problems caused by high dependence on imports of raw materials, urban centralization, and increasing energy consumption, the country's waste management policies (e.g., the volume-rate waste disposal system, the producer-responsibility recycling system, and the reduction of disposable products) have been recognized as a successful policy model by the international community. In the OECD Environmental Performance Review, which provides independent assessments of countries' progress toward their environmental policy objectives, the waste management of South Korea was among the highest in the OECD member countries [11].

However, the current waste management system is a recycling economy where resources are discarded after a few iterations of recycling, indicating that the country still has not overcome the limitations of the linear economy. Acknowledging these problems, the circular economy in South Korea requires an integrated strategy that encompasses a stable supply of raw materials and energy in addition to simple waste treatment and pollution management.

6.3 Goal and strategies of carbon neutrality

6.3.1 2050 Carbon neutrality goal

South Korea submitted the long-term low greenhouse gas emission development strategies (LEDS) to the United Nations in 2020. South Korea's LEDS vision was set out based on the principles of "contributing to global climate action," "laying the foundation for a sustainable and carbon-neutral society," and "actions at all levels." The key elements of the 2050 vision were laid out as follows, presenting the direction for policy, social, and technological innovation for a nationwide green transition [7].

❶ Expanding the use of clean power and hydrogen across all sectors
❷ Improving energy efficiency to a significant level
❸ Commercial deployment of carbon removal and other future technologies
❹ Scaling up the circular economy to improve industrial sustainability
❺ Enhancing carbon sinks

As an institutional foundation to pursue this carbon neutrality strategy, the Framework Act on Carbon Neutrality and Green Growth for coping with the climate crisis was enacted. South Korea became the 14th country to put the 2050 carbon neutrality vision into law and reflected its earnest determination for policy implementation by significantly enhancing its 2030 mid-term nationally determined contribution (NDC) target from the previous 26.3% reduction to a 40% reduction from the 2018 level [12] (Figure 6.5). South Korea is making utmost efforts to accomplish its stretched goal of a 4.17% annual reduction rate of NDC, compared to major countries (cf. EU 1.98%, the US 2.81%, the UK 2.81%, Japan 3.56%), by developing additional reduction measures and conducting related research.

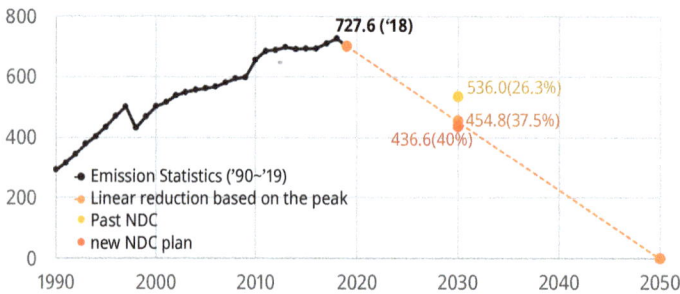

Figure 6.5: South Korea's NDC target (unit: million tons of CO_2eq).

To provide carbon reduction guidelines by sector, the Korean government announced scenarios of carbon neutrality [13]. The government developed the 2050 carbon neutrality scenario by setting up 10 technical working groups (general, energy conversion, industry, transportation, building, agricultural, livestock and fisheries, waste, carbon sinks, CCUS, hydrogen) composed of experts from 45 national research agencies recommended by 11 government departments, and through extensive discussions and reviews between related ministries. After disclosing the draft of the scenario, the final version of the scenario was completed through public hearing sessions to gather public opinions.

The carbon neutrality scenario provides two different scenarios, A and B, both of which aim for net zero emissions. While scenario A assumes a complete phase-out of fossil fuel power generation, scenario B assumes a small share of LNG-driven power generation together with wider use of technologies such as CCUS [13].

In scenario A, power generation driven by fossil fuels including LNG will be completely phased out, with no greenhouse gas (GHG) emissions emitted in the energy conversion sector, and GHG emissions in transportation and hydrogen will be minimized. As for some remaining carbon emissions in 2050, CCUS and carbon sinks such as forests will absorb and remove GHGs to achieve net zero emissions.

Scenario B assumes a phase-out of coal-fired power generation, but LNG power generation will partly remain. It is also assumed that there will be some internal combustion engines in operation using alternative fuels (e-fuels). In scenario B, however, carbon absorption and removal technologies will be significantly advanced and actively used so that net emissions will become zero, the same as in scenario A.

Carbon neutrality scenarios and reduction measures for each sector are listed in Table 6.1. Some of the key measures in the scenarios include abatement measures such as enhancing regulations and management, but new technologies will play a pivotal role. These carbon-neutral technologies are often still in the R&D stage, and therefore even more systematic R&D strategies are required to facilitate fast technology development and adoption.

Table 6.1: 2050 carbon neutrality scenarios in South Korea (unit: million tons of CO_2eq).

Sector	'18	Plan		Major reduction measures
		A	B	
Net emissions	686.3	0	0	
Energy conversion	269.6	0	20.7	– Reduction of thermal power generation – Renewable energy and hydrogen-based power generation
Industry	260.5	51.1	51.1	– (Steel) 100% replacement with hydrogen-reduced steel – (Petrochemicals) fuel and raw material conversion – (Others) energy efficiency improvement in the industry
Buildings	52.1	6.2	6.2	– Energy efficiency for new/remodeled buildings, facilities – Expansion of renewable energy, use of power plant waste heat
Transport	98.1	2.8	9.2	– Biofuels for shipping and aviation – Electricity/hydrogenation in the road sector
Agricultural, livestock	24.7	15.4	15.4	– Electricity/hydrogenation, biomass utilization – Livestock manure resource circulation, livestock management
Waste	17.1	4.4	4.4	– Restriction on disposable products, mandatory use of recycled materials, expansion of use of bioplastics – Recovery of landfill methane gas and energy utilization
Hydrogen	–	0	9	– (A) Supply with domestically produced green hydrogen – (B) Supply with part of domestically produced blue hydrogen
LULUCF*	–41.3	–25.3	–25.3	– Ecological restoration, expansion of new afforestation
CCUS	–	–55.1	–84.6	– (CCS) storage of CO_2 using domestic and overseas marine strata – (CCU) use of CO_2 by chemical conversion, etc.
DAC**	–	–	–7.4	– Capture of atmospheric CO_2 for e-fuel (plan B)

*LULUCF: land use, land-use change and forestry.
**DAC: direct air capture.

6.3.2 R&D strategies for carbon neutrality

The Korean government announced the "carbon neutral technology innovation strategy," and provides R&D support to secure the world's leading key technology for carbon neutrality [14] (Figure 6.6).

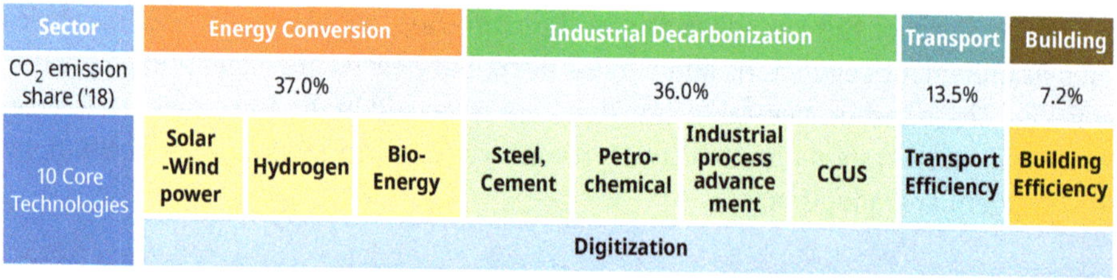

Sector	Energy Conversion			Industrial Decarbonization				Transport	Building
CO₂ emission share ('18)	37.0%			36.0%				13.5%	7.2%
10 Core Technologies	Solar-Wind power	Hydrogen	Bio-Energy	Steel, Cement	Petro-chemical	Industrial process advancement	CCUS	Transport Efficiency	Building Efficiency
	Digitization								

Figure 6.6: Carbon neutral technology innovation strategy.

With the vision of "driving 2050 Korea carbon neutrality by technological innovation," the strategy devised the top 10 key technologies and acquisition strategies with the urgent needs of the industrial sector reflected, in consideration of sectoral issues based on LEDS, level of contribution to GHG reduction, and the relevance to the main industries, as listed in Table 6.2 [14].

Table 6.2: The 10 core carbon neutrality technologies and R&D strategies.

Technologies		R&D strategies
① Solar and wind power	Solar	① Commercialization of next-generation high-efficiency technology ② Securing economic feasibility of water and offshore systems ③ Urban-type photovoltaic (ultra-light solar cell, etc.)
	wind	① Technological localization for core parts such as large blades ② Commercialization of large floating system ③ Securing system stability and operational reliability
② Hydrogen		① Low-cost and mass production technology of hydrogen ② Long-distance and large-capacity hydrogen storage and transport ③ High-efficiency, long-life fuel cell power generation system
③ Bio-energy		① Low-cost, high-quality, and mass-production technologies of bioenergy
④ Steel · cement		① Elimination of carbon-intensive emission process ② Technology for replacing fuels and raw materials
⑤ Petrochemical		① Biomass and recycled resource utilization ② Process electrification, low energy technology development

Table 6.2 (continued)

Technologies	R&D strategies
⑥ Process advancement	① High GWP process gases replacement · control · authentication technology ② Smart green device · process · factory advancement to improve industrial efficiency
⑦ Transport efficiency	① Next–generation battery and charging technology ② High durability and high output of fuel cell
⑧ Building efficiency	① Materials for zero-energy buildings ② Next-generation heating/cooling system and ICT convergence technology
⑨ Digitalization	① High efficiency of ICT devices and infrastructure ② Energy data integration and utilization technology ③ Next-generation power grid and large-capacity energy storage
⑩ CCUS	① Cost reduction through innovative material, efficiency, and enlargement ② Securing verification and stability through empirical research based on the domestic environment (emission source, storage, market demand)

On the basis of the strategy to acquire 10 key technologies, government ministries are working together to promote "full cycle running together" R&D activities through a two-track approach of "field-specific R&D" and "mid- to long-term basic R&D." To increase the penetration of new technologies, the regulatory sandbox is to be expanded and support for demonstration and commercialization is to be strengthened. Further, support for the full-cycle growth of carbon-neutral companies will be provided through support for start-ups, linkage with public demand purchases, and more support for green finance.

The Ministry of Trade, Industry and Energy (MOTIE) announced the "carbon-neutral R&D strategy for industry and energy" in November 2021, and selected 17 R&D focus areas for carbon neutrality and derived key technologies for each area with the aim to fulfill the 2030 NDC goal and achieve 2050 carbon neutrality neutrality [15].

6.3.3 R&D investment trends for carbon neutrality

To implement the strategies for carbon-neutral R&D, each ministry is preparing for related R&D projects. These projects include commercialization-oriented development projects for carbon-neutral key technologies by sector, energy demand management and high-efficient energy development, CCUS, industrial process innovation, support for small- and medium-sized enterprises, and manpower training to develop next-generation technologies with high potential to contribute to carbon neutrality (Figure 6.7).

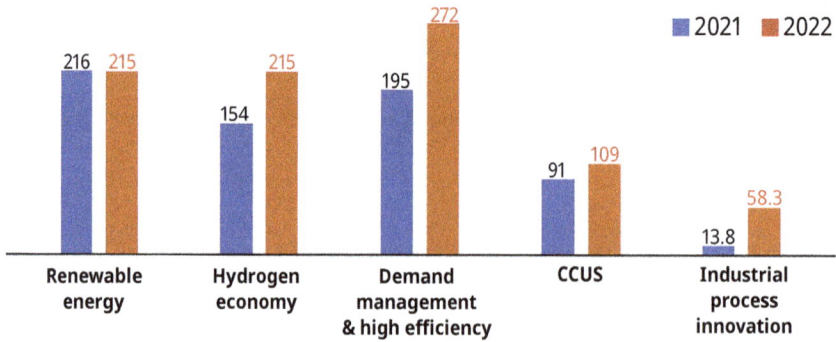

Figure 6.7: Carbon-neutral R&D budget in major sectors (unit: USD 1 million).

As shown in Figure 6.8, the government's investment in carbon-neutral R&D is on the rise at a growth rate of 15.5% annually. The carbon-neutral R&D investment in 2022 amounts to $1.65 billion (cf. $1.4 billion in 2021), but if the investment for digital transformation research that falls under the carbon-neutral technology is included, the carbon-neutral R&D investment will increase up to $3 billion [16, 17].

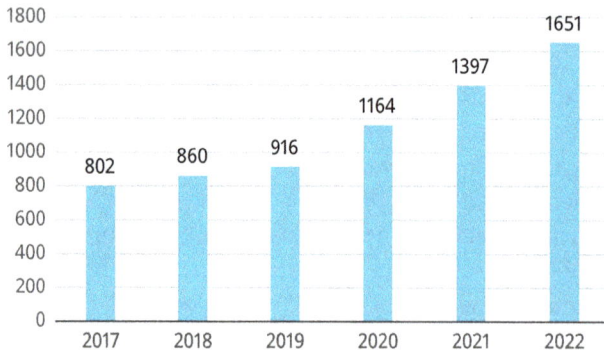

Figure 6.8: The government's investment in carbon-neutral R&D (unit: USD 1 million).

Each ministry has formulated a variety of strategies for carbon neutrality technologies and is promoting R&D projects accordingly. The Ministry of Science and ICT (MSIT) announced the carbon neutral technology innovation strategy and CCU technology innovation roadmap, and the Ministry of Trade, Industry and Energy (MOTIE) announced the carbon neutrality vision and strategy for industry and energy, the first hydrogen economy implementation masterplan in 2021. The Ministry of Oceans and Fisheries (MOF) announced the 2050 carbon-neutral roadmap in the oceans and fisheries sector, and the Ministry of Land, Infrastructure and Transport (MOLIT) announced the 2050 carbon-neutral roadmap in land and transport sector. As seen in Table 6.3, Korean gov-

Table 6.3: Current status of large-scale net-zero R&D projects.

Project	Ministry	Main fields	Period
Net-zero innovative technology development	MSIT	Solar power, wind power, industrial process efficiency, building efficiency, and secondary batteries	'23–'30
Net-zero industrial core technology development	MOTIE	Steel, cement, petrochemical, automobile, shipbuilding, electrical and electronic industries	'23–'30
Donghae CCS integration demonstration	MOTIE	Construction and utilization of CCS facilities utilizing the depleted gas field in the East Sea	'23–'30
Low-carbon ecosystem for SMEs and venture companies	MSS*	Low-carbon technology development tailored to small- and medium-sized businesses	'23–'30
Digital-based carbon-neutral city technology	Joint ministries	City-level demonstration of carbon-neutral technology, AI/data integration platform	'23–'29
Core technology for building a low-carbon hydrogen supply chain	Joint ministries	Green hydrogen production, overseas hydrogen introduction technology development, etc.	Undefined
3050 CCU R&D and integrated demonstration	Joint ministries	Advancement of R&D and demonstration for 14 strategic products	'24–'30
Carbon-neutral university base center	MOTIE	Net-zero innovation platform within the university, human resources development	Undefined

*MSS: Ministry of SMEs and Startups

ernment ministries are planning or applying for a large-scale R&D project to implement the aforementioned strategies [16].

South Korea operates 25 government-funded research institutes and conducts a variety of studies on carbon neutrality in an effort to secure state-of-the-art science and technology. Korea Research Institute of Chemical Technology (KRICT), Korea Institute of Science and Technology (KIST), Korea Institute of Geoscience and Mineral Resources (KIGAM), Korea Institute of Energy Research (KIER), Korea Institute of Materials Science (KIMS), Korea Institute of Machinery and Materials (KIMM), and many other institutes are carrying out carbon neutrality-related research. Here, we introduce the research case of KRICT, which is responsible for developing numerous technologies of the carbon-neutral technology innovation strategy (Table 6.4).

Table 6.4: The carbon neutrality research of KRICT.

---

1. About KRICT (Korea Research Institute of Chemical Technology)

KRICT has played a pivotal role in the advancement of the nation's chemical industry through the development of technologies in the chemical and relevant convergence fields, transferring chemical technology to industries, fostering experts and professionals, and providing a wide range of support for chemical-related infrastructure.

In KRICT, five research divisions are in operation including chemical processes, advanced materials, therapeutics and biotechnology, specialty and bio-based chemicals, and chemical platform, and a majority of these studies are focused on environmental-friendly research such as climate response, energy independence, and resource circulation. About 620 researchers are conducting research with an annual budget of approximately $218 million.

---

2. Summary of key research related to the circular economy in KRICT

KRICT has extensive R&D experience, expertise, and infrastructure from the source technologies in the energy and environment fields. As for the resource circulation technology, technologies to achieve carbon neutrality in the petrochemical industry are under development such as low-energy process technology, CO_2 chemical conversion technology, byproduct gas conversion technology, waste plastic depolymerization technology, and biomass utilization technology (Figure 6.9). With regard to renewable energy-related technology, the institute is developing core technologies for fuel cells and secondary batteries that can utilize sources of renewable energy such as solar power, hydrogen, and bioenergy.

KRICT has secured precommercialization technology by completing a demonstration of technology to convert CO_2 to methanol at a capacity of 10 tons/day in 2015. The technology for capture of CO_2 from power plant exhaust gas was studied through a demonstration plant with a capacity of 10 tons CO_2/day. It exhibited a 30% energy reduction compared to the existing dry process.

Furthermore, the institute has developed new process technology for NMTO (naphtha and methanol to olefin) that can save about 30% of energy consumption compared to existing pyrolysis technology of raw materials of base chemicals, and currently a TRL6 stage empirical study is in progress. KRICT has secured the source technology to convert CO_2 to a high value-added material without a CO_2 capture process and a catalyst source technology that can produce styrene from waste Styrofoam or obtain high-value monomers from a colored PET at low temperature/eco-friendly conditions.

Moreover, KRICT is showing the world's best performance in the field of perovskite solar cells. Through continuous R&D effort in solvent engineering, it achieved the world's highest efficiency of 25.2% at a power generation efficiency of $0.1cm^2$ unit devices (NREL, 2019). It also has secured a roll-to-roll production process technology to achieve a high level of compatibility.

In addition, the institute has secured the source technology for secondary battery materials with high stability and high efficiency, and continues its research on green hydrogen production, hydrogen production technology without CO_2 emissions, and the development of safe hydrogen storage.

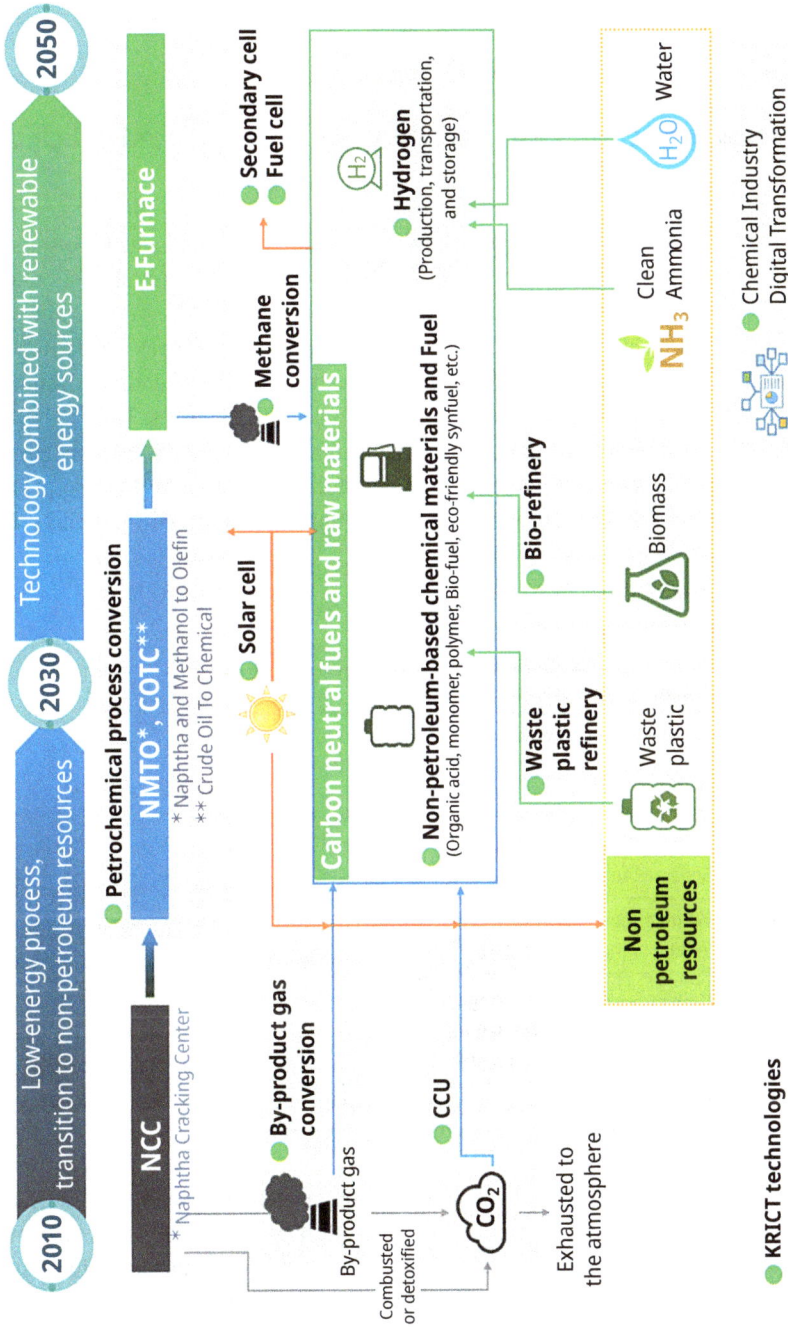

Figure 6.9: KRICT's carbon–neutral technology blueprint.

6.4 South Korea's laws and system for the circular economy

6.4.1 Framework Act on Resource Circulation and related policies

The Framework Act on Resource Circulation was enacted in 2018 with the purpose of facilitating circular utilization and appropriate treatment of generated wastes, beyond promoting resource saving and recycling. The first master plan for resource circulation is mainly focused on the disposal phase, such as reducing the generation of wastes and the amount of landfill and increasing recycling. More work needs to be done in this area to identify the efficiency of raw material usage, the circulation of material flow, and the extent of innovation and national competitiveness [18].

Table 6.5 exhibits the key waste utilization-related systems in South Korea. There are a variety of waste-related systems, in the form of direct regulation that prohibits or obliges certain actions and indirect regulation that uses financial incentives to achieve policy goals.

Table 6.5: Korea waste treatment and utilization system.

Division	Treatment and utilization system
Direct regulation	Suppression of excessive and disposable packaging (2022)
	Packaging material structure evaluation system (2019)
	Recycling environmental assessment (2016)
	Resource circulation performance management system (2014)
	Environmental guarantee system (2008)
	Separate emission labeling system (2003)
Indirect regulation	Waste disposal fee system (2018)
	Voluntary agreement system (2008)
	Producer responsible recycling system (2003)
	Garbage volume system (1995)
	Waste charge system (1993)
	Empty container deposit system (1985)
etc.	Recycling resources recognition system (2018)

In 2020, the government established the great transition promotion plan for its resource circulation policy to revise and supplement the existing key action plans since waste collection has been hampered by the persistent sluggish recycling market with fast-growing waste generation after the COVID-19 pandemic. Its implementation plan is composed of promotion plans for the entire life cycle of resource circulation by dividing the plan into ①generation phase, ②discharge/collection phase, ③selection/recycle phase, ④final treatment phase, ⑤inspection and management of the implementation.

However, since the current Framework Act on Resource Circulation and the great transition promotion plan for resource circulation policy are confined to the waste sector, we need a circular economy system with sustainability taken into account to facilitate the efficient use and circulation of resources in the life-cycle of production–distribution–consumption.

6.4.2 Establishment of Korean (K)-circular economy implementation plan

As wastes have been used as a key element to achieving carbon neutrality, it is necessary to make even broader policy considerations for the life cycle of wastes. In 2021, the Korean government announced the Korean (K)-circular economy implementation plans for carbon neutrality to achieve complete circulation of waste resources to significantly reduce GHG emissions in the industrial sector and make use of new growth engines.

The following are the four pillars of the implementation plan: ① strengthening the resource circulation in the production and distribution phase, ② promoting eco-friendly consumption, ③ expanding waste resource recycling, and ④ a stable waste resource treatment system. It has set performance indicators that were previously confined to waste treatment as the main and subindicators for each step of the life-cycle resource circulation management and the expansion of the circular economy implementation. The application methods and targets will be determined by gathering opinions from relevant ministries and industries [19].

6.5 Efforts of Korean businesses for the circular economy

In addition to the carbon border tax, the European Union (EU) is reviewing a bill to ban financial institutions from investing in carbon-intensive industries. Thus, it is likely that the businesses could face increasing pressure to reduce carbon emissions; otherwise, the European investors could withdraw their investments. As a result, Korean companies with a high share of overseas exports are making their hardest effort to respond to carbon neutrality and a circular economy.

6.5.1 Technology development for carbon neutrality and resource circulation

Businesses are taking various actions to obtain technologies for carbon neutrality and resource circulation. They are looking to seek a diverse range of technologies from technology for chemical recycling of plastics to waste management.

Plastic pyrolysis technology, in particular, is gaining tremendous attention from major domestic petrochemical companies. Hyundai Chemical, SK Geocentric, LG Chem, and Lotte Chemical are either building production facilities or have started production. Hyundai Chemical plans to initiate eco-friendly petroleum product production using plastic pyrolysis oil from June 2022 and has already acquired ISCC PLUS certification for this purpose.

Companies in various sectors such as petrochemical, construction and steel are showing interest in CCU Technologies, LG Chem, Lotte Chemical, Hyundai Steel, and Samsung ENG are one of those companies. Among them, DL E&C has commercialized a carbon capture plant, the first case in Korea, and is capable of capturing one million tons of carbon per year. As the first domestic construction company, it has successfully entered the global market and is in preparation for building the first carbon-negative plant in Korea.

In addition to these, the company is pushing ahead with the White Bio business that uses nonedible organic wastes, renewable energy projects, and waste recycling businesses. Hyundai Motor plans to promote the electrification of all vehicles for sales with the target to achieve carbon neutrality by 2045 and accelerate the transition to renewable energy and green hydrogen. The company has set a goal to increase the share of electrified models up to 80% of the vehicles sold worldwide by 2040. Samsung Electronics has enhanced the sustainability of the semiconductor business by using renewable energies at overseas sites and promoting GHG projects. As a result, all semiconductor business sites of Samsung Electronics are making progress in carbon neutrality such as obtaining the first carbon, water, and waste reduction certification of carbon trust in the industry.

To secure new technologies, some companies use M&A activities. Hanwha Solutions acquired "RES Méditerranée SAS," a French renewable energy company, for $860 million. SK Eco Plant purchased four waste treatment companies for $367 million after the acquisition of EMC Holdings for $873 million in 2020. In February 2022, the company acquired electric and electronic waste (e-waste) company to embark on building an e-waste circular economy. The number 1 domestic cement company, Ssangyong C&E, has also acquired four waste treatment companies.

6.5.2 Corporate strategies and investment plans

To achieve carbon neutrality, many companies are getting ready for long-term commitment, and announcing corporate strategies. POSCO, criticized for being the largest GHG emitter in Korea, announced the 2030 mid- to long-term growth strategy in 2021 and revealed that it would promote eco-friendly projects such as green steel development without carbon emissions.

LG Electronics set a goal to produce eco-friendly products based on three elements: energy, human, and resources. For the energy element, the company plans to set a five-year technology roadmap for each product line and achieve detailed targets to improve product energy efficiency, and minimize environmental impact through continuous investment in R&D.

Hyundai Motor Group's major affiliates are implementing strategies to become a carbon-free brand with the aim of 2045 carbon neutrality. Hyundai Motor has a plan to reduce carbon emissions from vehicle operation, supply chains, and business sites by 75% compared to the 2019 level and achieve zero emissions by 2045 with CCUS technology.

Hanwha Group launched the ESG Committee in 2021 and is committed to addressing environmental issues and promoting sustainable businesses by expanding eco-friendly technologies such as solar energy, wind energy, green hydrogen, and plastic recycling.

These companies are also working to secure sufficient budgets to achieve targets for carbon neutrality and implement the plan. The five companies, including Hanwha Corporation, Hanwha Solutions, Hanwha Aerospace, Hanwha Energy, and Hanwha Total Energies, secured funds by issuing $1.06 billion of ESG bonds in 2021. Using these funds, they plan to invest in the production line expansion for solar cells and modules, prevention of air pollution, energy efficiency improvement, and the construction of eco-friendly buildings. In the next five coming years, Hanwha Group will invest $3.7 billion in the energy sector including solar and wind power and a total of $2.6 billion in the carbon-neutral sector.

SK Group announced its plans to invest $216 billion in future growth industries by 2026. In specific, $58.5 billion will be invested in eco-friendly industries such as electric vehicle (EV) batteries and battery materials, hydrogen, wind power, and renewable energy. In January of last year, SK Corporation and SK E&S made a joint investment of approximately $1.57 billion in Plug Power, a hydrogen fuel cell company.

Kumho Petrochemical announced an investment plan to build an ESG-leading business system and shore up eco-friendly businesses. The company plans to establish sustainable strategies with ESG management and carbon-neutral growth as a basis, and invest roughly $2.36 billion in related businesses in the next five years.

6.5.3 Networking activities

Given the short preparation period for carbon neutrality, companies in Korea are working hard to expand synergies based on various networks. For instance, LG Chem held a business agreement ceremony for building a virtuous cycle for PVC waste wallpaper resources with local governments and companies. The three entities are planning to lay an institutional foundation to use PVC waste wallpaper as a sustainable resource.

SK runs the Green Campus where six affiliates that promote eco-friendly projects take part. Green Campus is a "shared infrastructure" for six affiliates with eco-friendly businesses to discover and develop new future businesses through close collaboration between them. Moreover, SK also plans to establish the SK Green Techno Campus that musters R&D manpower and capabilities in the field of eco-friendly business. The campus, as an R&D infrastructure dedicated to developing new green business technologies, will help the company swiftly meet the demand for eco-friendly technology development and create synergy by allocating the right technological infrastructure and manpower of related companies in the right place.

Furthermore, it is actively participating in the carbon-neutral and circular economy-related consultative body run by the governments, government-funded companies, and local governments, to deliver corporate opinions in the policy-making process. To facilitate the adoption of carbon-neutral technology, the businesses seek a stable supply of raw materials such as waste plastics, biomass, and byproduct gas, and hope to see incentive systems or regulations to compensate for the higher price compared to existing commodity products. In the process of adopting new technologies, businesses want clear standards and eased licensing criteria that recognize GHG reduction so that carbon-neutral products can be quickly rolled out to the market.

To address this issue, related businesses and research institutes have put together several consultative bodies to voice their ideas, such as the Carbon Neutrality Chemical Technology Research Council led by KRICT with 18 private companies and the Energy Alliance and Green Ammonia Council for Carbon Neutrality.

6.6 The will of citizens and society

6.6.1 Public awareness of the circular economy

The Korean government surveyed the public awareness of carbon neutrality involving 1,600 people in 2021. As a result, most of the respondents thought that the abnormal global weather and natural disasters are serious (96.3%), and that climate change is having an impact on their daily life (91.9%) [20] (Figure 6.10).

(unit: % / n=1,600)

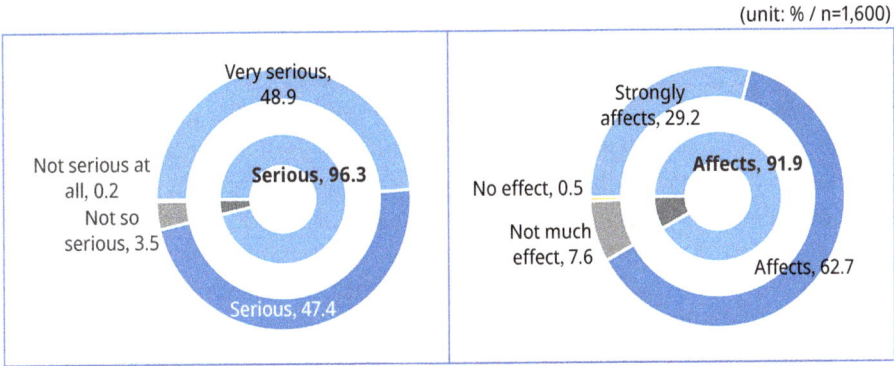

Figure 6.10: The public awareness of climate change (severity, sensitivity).

In the survey questioning citizens' willingness to participate in carbon-neutral activities, "reducing the use of disposable products and food waste" received the highest response (92.5%), followed by "increasing the use of public transportation and bicycles" at 88.3% and "using less electricity by improving behavior" at 87.6%, "reinforcing insulation and installing triple glazing windows to improve house energy efficiency" at 84.9%, "purchasing and replacing electric/hydrogen vehicles if needed" at 77.7%, "participating in resident participatory renewable energy projects such as village-wide solar power systems" at 73.0%.

To incorporate public opinion in the 2050 Carbon Neutrality Plan, the Carbon Neutral Citizens Council joined by 513 citizens was launched in 2021. The Citizens Council organized for citizens aged 15 and over based on region, age, and gender participated in the public hearing session to speak up their opinions after learning about carbon neutrality through citizen carbon classes, distribution of learning materials, and e-learning. They showed strong interest and determination of citizens for carbon neutrality with 90.4% of participation rate on the first day of the hearing session and 89.3% on the second day. The 2030 enhanced update of NDC and the carbon neutrality scenario were published after reflecting public opinions [12, 13].

6.6.2 Voluntary actions for carbon neutrality

The Korean people are taking various civic actions to achieve carbon neutrality. For example, there are certain groups organized by the representatives of generations. One of them is the 60 + Climate Action group of 524 senior citizens over 60, and the other is the Baby Climate Action Group where 62 children under age 10 took part.

The 60 + Climate Action group called on senior citizens to act on climate change, stressing that they are the "causes" of the climate crisis and that "it is their duty and the right of old people to pass on a better environment than what was inherited" in a declaration. This action group is regarded as one of the "gray-green" movements that are spreading in the countries like the US and Europe.

The "Baby Climate Action Group" is composed of 39 children under age 5, 22 children between the ages of 6 and 10, and one newborn baby of 20 weeks of age. They filed a constitutional complaint, claiming that the Enforcement Decree of the Framework Act on Carbon Neutrality for coping with climate crisis violated the basic rights of future generations. They argue that even when the 2030 NDC goal presented in the Carbon Neutrality Framework Act is achieved, the right of future generations to pursue happiness will be infringed. They concluded that it was unconstitutional to pass on the burden of GHG reduction to future generations.

Furthermore, solar power generation cooperatives have been formed to help build solar systems, and recyclable products are widely collected and recycled through local resource circulation campaigns. The MZ generation, dubbed the young generation in Korea, prefers to consume environmental-friendly products and does not mind the laborious or more expensive nature of purchasing and using them. Such tendency triggers companies' interest in eco-friendly activities, driving more attention of these companies to sustainability.

6.7 Future plans

The peak of GHG emissions in Korea is estimated to be reached in 2018, meaning that the country only has 32 years to achieve carbon neutrality by 2050, which makes Korea's carbon neutrality target even more challenging compared to other advanced countries. However, Korea has a much higher proportion of manufacturing industries that consumes a lot of energy, compared to other major advanced countries, and the share of renewable energy is about 7% in 2020, the lowest among the OECD countries.

The government is working hard to formulate strategies to ensure that no one is harmed by the introduction of carbon neutrality, reflecting the opinions of the public and the experts in the relevant field as much as possible, and securing budget to facilitate science and technology development as a way to overcome geographical limitations. What remains now is faithfully implementing the national carbon neutrality strategy devised after much deliberation and achieving carbon neutrality.

The carbon-neutral policy would be further revised and complemented by the newly established administration. Recently, the MOTIE announced the 2030 electricity supply plan through the 10th Basic Plan for Electricity Supply and Demand, increasing the share of nuclear power generation in 2030 from 23.9% to 32.8% and lowering the share of renewable energy from 32.2% to 21.5% [21]. In accordance with the new administration's nuclear-friendly policy direction, the forecast was adjusted to reflect the continuous operation of existing nuclear power plants and the generation capacity of new nuclear power plants. In this projection, the share of renewable energy power generation was adjusted downward considering residents' acceptance and feasibility.

In addition, the proportion of coal-fired power generation was slightly reduced from 21.8% to 21.2%, and the share of noncarbon power source generation such as hydrogen/ammonia cofiring power generation decreased from 3.6% to 2.3%. The projections for the share of hydrogen/ammonia cofiring generation have taken into account the actual level of fuel supply and corporate intentions, while the forecast for the share of coal-fired power generation reflected additional reduction plans such as shutdown plans and capping operation rates.

By reflecting these changes, the existing carbon-neutral strategy will be revised to some extent. Nevertheless, South Korea will muster strength from the government, industry, and citizens to continuously and actively respond to the global issue of carbon neutrality.

References

[1]   Kim, E.A. and Min, B.K., Analysis of strategies and case studies on regional circular economies. National Assembly Futures Institute, 2020.12, 20–06.

[2]   Geissdoerfer, M., Savaget, P., Bocken, Nancy M.P. and Hultink, E.J., The circular economy e A new sustainability paradigm?. Journal of Cleaner Production, 2017, 143, 757–768.

[3]   European Commission. Closing the loop – an EU action plan for the circular economy. COM, 2015, 614 final.

[4]   Kirchherr, J., Reike, D. and Hekkert, M., Conceptualizing the circular economy: An analysis of 114 definitions. Resources, Conservation & Recycling, 2017, 127, 221–232.

[5]   Korea Research Institute for Human Settlements. KRIHS issue report, A plan to implement a safe national land in response to climate change disasters. 2022. 04.

[6]   Korea Meteorological Administration. Korea Climate Change Assessment Report 2020. 2020.07.

[7]   The Government of the Republic of Korea. 2050 Carbon neutral strategy of the republic of Korea. 2020.10.

[8]   Ministry of Trade, Industry and Energy (a). Carbon–neutral R&D strategy for Industry and Energy. 2021.11.

[9]   Statistics Korea. KOSIS, Korean statistical information service. 2022. (Accessed Sep 30, 2022, https://kosis.kr/index/index.do)

[10]   Green Technology Center. A study on technology level assessment and key promising–area selection in fields of climate technology. 2020.12.

[11]   OECD. Environmental country reviews_Korea. 2017.

[12]   Carbon Neutrality Commission (a). The 2030 enhanced update of NDC. 2021.10.

[13]   Carbon Neutrality Commission (b). 2050 Carbon Neutrality Scenario. 2021.10.

[14]   Ministry of Science and ICT. Carbon Neutral Technology Innovation Promotion Strategy. 2021.09.

[15]   Ministry of Trade, Industry and Energy (b). 2050 Carbon Neutral Energy Technology Roadmap. 2021.12.

[16]   Presidential advisory council on education, science & technology. Mission–oriented, Carbon Neutral R&D Promotion Plan. 2022.02.

[17]   Ministers' Meeting on Science and Technology. Carbon Neutral R&D Investment Strategy Towards 2050 Carbon Neutrality. 2021.03.

[18] Bae, J.S.,. Status of domestic circular economy and policy implications – focusing on economic incentives. KDI, 2021.10.

[19] Ministry of Trade, Industry and Energy. Ministry of Environment. Establishment of Korean(K)–Circular Economy Implementation Plan. 2021.12.

[20] Ministry of Culture, Sports and Tourism. Report on Public Perception on the Promotion of 2050 Carbon Neutrality. 2021.11.

[21] Ministry of Trade, Industry and Energy (c). The 10th Basic Plan for Electricity Supply and Demand. 2022.08.

Index

https://doi.org/10.1515/9783110767179-007

Also of interest

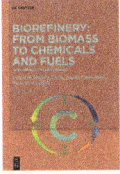

Biorefinery: From Biomass to Chemicals and Fuels.
Towards Circular Economy
Aresta, Dibenedetto, Dumeignil (Eds.), 2022
ISBN 978-3-11-070536-2, e-ISBN (PDF) 978-3-11-070538-6,
e-ISBN (EPUB) 978-3-11-070541-6

Circular Plastics Technologies
Knauer, 2024
ISBN 978-1-5015-2328-1, e-ISBN (PDF) 978-1-5015-1561-3,
e-ISBN (EPUB) 978-1-5015-1562-0

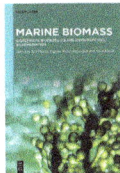

Marine Biomass.
Biorefinery, Bioproducts and Environmental Bioremediation
Kapoor, Rafatullah, Ismail (Eds.), 2024
ISBN 978-3-11-135358-6, e-ISBN (PDF) 978-3-11-135395-1,
e-ISBN (EPUB) 978-3-11-135406-4

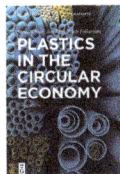

Plastics in the Circular Economy
Voet, Jager, Folkersma, 2024
ISBN 978-3-11-120029-3, e-ISBN (PDF) 978-3-11-120144-3,
e-ISBN (EPUB) 978-3-11-120247-1

Bio Products
Green Materials for an Emerging Circular and Sustainable Economy
Vijayendran (Ed.), 2023
ISBN 978-3-11-079121-1, e-ISBN (PDF) 978-3-11-079122-8,
e-ISBN (EPUB) 978-3-11-079148-8